ACTIVITIES

Maths Problem Solving

AGES 5-7

LOUISE CARRUTHERS

Author
Louise Carruthers

Editor
Sally Gray

Assistant Editor
Linda Mellor

Series Designers
Anthony Long and
Joy Monkhouse

Designers
Allison Parry
Catherine Mason

Illustrations
Louise Gardner

Photographs
Derek Cooknell

The publishers would like to thank:

Gerry Bailey and the staff and pupils at
**Clapham Terrace First School,
Leamington Spa**.

Published by Scholastic Ltd
Book End
Range Road
Witney
Oxfordshire OX29 0YD

www.scholastic.co.uk

Text © 2006 Louise Carruthers
© 2006 Scholastic Ltd

Designed using Adobe Indesign

Printed by by Bell & Bain Ltd, Glasgow

10 11 12 13 14 15 16   2 3 5

British Library Cataloguing-in-Publication Data

A catalogue record for this book is available from
the British Library.

ISBN 0-439-96556-X
ISBN 978-0439-96556-9

MIX
Paper from
responsible sources
FSC® C007785

Contents

Introduction

Why teach problem solving?

'The ability to solve problems is at the heart of mathematics.'

Cockcroft Report (DES 1982)

There are many reasons for including problem-solving as an integral part of the primary mathematics curriculum. Well-planned problem-solving activities provide opportunities for children to:

- apply their mathematical knowledge and skills creatively and flexibly in a wide range of situations;
- develop their reasoning and communication skills;
- work co-operatively;
- develop confidence and independence in their own mathematical ability.

Classification of problems

There are several types of mathematical problem. The DfES publication *Problem solving* (DfES, reference: 0248-2004 G)

classifies problems into five main categories. They are as follows:

1. Finding all the possibilities
2. Logic problems
3. Finding rules and describing patterns
4. Diagram problems and visual puzzles
5. Word problems.

The activities in this book are divided into four chapters. Each chapter focuses on a particular type of problem (1 to 4 above) and the relevant problem-solving strategies. Further information about each of the different types of problem is given at the beginning of each chapter. Word problems (5) are not included in this book. Many examples of word problems can be found in the DfES's *Framework for teaching mathematics*.

A creative approach to problem solving

When solving open-ended mathematical problems, children should be encouraged to explore different strategies and make decisions about how to record outcomes. In order to do this, they are required to think flexibly and creatively. Many problems have more than one possible answer and can be solved using different approaches. It is important that children are confident to take risks, to question their ideas and to adapt their approaches as they work.

Each mathematical problem in this book is presented creatively in an imaginative or relevant context to engage the children and motivate them to solve it. Popular children's fiction, artwork, and cultural traditions are some of the starting points used to introduce the problem-solving activities.

The activities in this book are designed to be used flexibly and the creative contexts can easily be adapted to reflect children's particular interests or to make cross-curricular links. For example, instead of arranging tarts on a plate ('Queen of Hearts', pages 86-87) the children could put spots on dogs or shells on sandcastles! Some of the problems can be personalised by using the children's names for the characters in a problem (such as in 'Keep off!', pages 62-63).

How to use the activities in this book

Each problem is intended to last for one lesson. Sometimes the lesson may need to be longer than the hour set aside for the daily numeracy lesson, to allow the children sufficient time to complete the activity.

Many of the activities in this book encourage children to devise their own strategies for solving problems and recording outcomes. Different groups of children may respond to the problems in quite diverse ways. Consequently, it may often be necessary to adapt the suggested plan of an activity to meet the needs of a particular group of learners.

In general the problems at the beginning of each chapter are the easiest and they get progressively more difficult. Each of the problem-solving lessons is structured in the same way:

Setting the context

The problem is presented creatively to engage all the children and appeal to different learning styles. The children are encouraged to think about how they are going to solve the problem, what equipment they will use and how they will record what they do.

Solving the problem

The children work independently, in pairs or in small groups to solve the problem. While they are working, they are encouraged to talk about what they are doing. Targeted support is provided where necessary.

Drawing together

Different groups are given the opportunity to report back to the rest of the class. They are asked to describe the strategies they used to solve the problem and explain how they

recorded their work. The children are helped to evaluate the different strategies which have been used and, where appropriate, they are shown how these strategies can be improved.

It is important that in every lesson the problem is displayed prominently so that children can refer to it during the lesson.

Organisation

Most of the activities are suitable for collaborative work. Organising the children to work in pairs or small groups will encourage co-operation and communication and give children the confidence to be creative and flexible in their thinking.

Differentiation

While a problem should have a degree of challenge, it should not be too hard – the children need to feel that they have a chance of solving it. Problems can easily be adapted to suit children of all levels by simply changing a number or giving more direction about how to solve the problem. At the end of each problem, there are suggestions for further differentiating the activities for children who need more support, as well as ideas for how to extend children who have achieved the lesson objectives.

Further ideas

The activities involve a lot of direct teaching and demonstration. At the end of each activity one or more further ideas are suggested. These problems are generally the same problem, set in a different context. It is recommended that the children are set one or more of these follow-up problems to reinforce and consolidate the learning that has taken place during the structured lesson.

Developing a positive approach to problem solving

The role of the teacher

The primary role of the teacher in the creative problem-solving process is to equip children with appropriate knowledge, skills and attitudes, in order that the children may feel empowered to tackle any problem. It is important that all teachers:

● present problems creatively to engage all children

● make children aware that there are different types of problem and that there are many ways of working out solutions

● show children how they can use and apply the mathematical skills and knowledge they already have to solve problems

● teach children specific problem-solving skills and systematic approaches to recording outcomes

● develop children's reasoning and communication skills by allowing time for discussion of ideas and asking carefully planned questions

● offer support and encouragement to help children 'keep going' with more challenging problems.

Scaffolding learning

It is imperative that teachers offer appropriate support and intervention at each stage of the problem-solving process. A crucial aspect of the teacher's role in scaffolding learning is to ask the children carefully planned leading questions. The following questions could be used to guide children through each stage of the problem-solving process:

Understanding the problem

● What is the problem asking you to find out?

● Have you seen a problem like this before?

● Can you tell me the problem in your own words?

● What do you know that might help you solve this problem?

Getting started

● Can you guess what the answer might be?

● Are there any rules that you must follow? What equipment could you use to help you?

Solving the problem

● What are you doing?

● Why have you done this?

● Can you tell me what you have done so far?

● What do you think would happen if...?

● Have you tried...?

● Can you think of an easier way of recording what you are doing?

● What do you notice if...?

Drawing together

● What strategy did you use to get your answer?

● Can you think of another method you could have used?

● How did you check your answer?

● Which group do you think had the best strategy for solving the problem? Why?

● How did you record your solutions?

● Do you think you could have solved the problem more quickly if you had recorded the outcomes in a different way?

● Can you make up a similar problem?

Chapter One

Finding all the possibilities

Many problems have more than one possible solution. Problems that require children to find all the possibilities can be solved by trial and error, but this is not an efficient approach – answers can easily be repeated and it is difficult to know when all the possibilities have been found. Children need to be taught more effective strategies for finding all the possibilities and they need to be shown how these strategies can be applied systematically.

The activities in this chapter introduce children to the need for systematic working when finding all the possibilities. **The two main strategies of organising and classifying data that they are taught are:**

● how to sort or reorder data that has been collected to make it easier to see which answers have not been included. For example, in 'Colourful caterpillars' (see pages 24-25) the children are told to record each answer on a separate piece of paper. They are then shown how to sort the answers into four different sets to make it easier to identify any answers they may have missed out;

● to implement a system for finding all the possibilities as data is being collected. For example, in 'Red packet' (see pages 10-11) the children record the possibilities in an ordered list starting with the smallest number.

Children's ability to work systematically will progress at different rates. It is necessary for some children to see the process of working systematically modelled and explained many times before they are ready to apply these strategies independently. Other children may understand more clearly the need for systematic working and these children may need less direction and help in order to apply different systems for recording all the possibilities. The problems in this chapter can easily be adapted to suit children of all abilities.

The Rainbow Fish

Setting the context

This lesson uses *The Rainbow Fish* by Marcus Pfister (North-South Books) as a starting point. The Rainbow Fish was the most beautiful fish in the sea but he had no friends because he would not share his shiny scales with the other fish.

Problem

How many different ways can the Rainbow Fish share his scales with his friend the little blue fish?

Objectives

To solve mathematical problems and puzzles. To know addition facts for numbers up to 10. To begin to have a system for finding all possibilities.

You will need

The Rainbow Fish by Marcus Pfister (North-South Books); shiny paper; scissors; double-sided sticky tape; photocopiable page 28; water-related music, such as Handel's *Water Music*, Debussy's *Clair de Lune* or Michael Nyman's soundtrack to *The Piano*.

Preparation

Cut out six shiny scales. Read *The Rainbow Fish* before starting this activity.

Solving the problem

● Organise the children so that they are sitting in a large circle. Pick a child to be the 'Rainbow Fish' and stick the six shiny scales onto his top using double-sided sticky tape. Pick a second child to be the little blue fish. Remind the children how the Rainbow Fish makes friends with the other fish in the sea by sharing his beautiful scales with them.

● Explain that you are going to play some music and that you would like the two fish to swim around in the 'sea'. Tell the Rainbow fish that every time the music stops he must give one of his scales to the little blue fish. Each time the Rainbow Fish gives one of his scales to the little blue fish the rest of the class needs to help you count how many scales each fish now has, saying how many scales there are altogether. Continue until the little blue fish has all six shiny scales.

● Gather the class around the whiteboard. Read the problem to the class and explain that you would like them to help you find and record all the different ways it is possible to split the six scales between the two fish.

● Choose a different child to be the Rainbow Fish and wear the six shiny scales. Let the Rainbow Fish choose a friend to share her scales with.

● Draw this simple table on the board. Explain that the table shows how many shiny scales each fish has.

Rainbow Fish	Little blue fish
6 shiny scales	0 shiny scales

● Tell the Rainbow Fish to give one of his scales to his friend. Pick a child to record how many scales each fish has now in the table (5,1). Repeat this process until the Rainbow Fish has no scales left: (4,2) (3,3) (2,4) (1,5) (0,6).

● Ask: *How many different ways did we find to split the six scales between the two fish? Do you think we have listed all the possibilities? How do you know?* Explain to the children that because you have worked systematically, starting with the largest number and counting only one scale at a time, you can be certain that you have found all the possible ways that the two fish can share six scales.

● Organise the children to work in pairs. Give each pair a copy of photocopiable page 28, some shiny paper, scissors and a pencil.

● Challenge the children to investigate the number of ways the Rainbow Fish can share seven scales with his friend. Invite the children to record all the different solutions they can find in the table.

● Circulate around each of the groups while the children are working. Talk to the children about what they are doing and observe how the children tackle the problem. Do they apply a hit and miss approach to finding all the possible answers? Or do any of them use the system modelled above, recording all the possibilities in an ordered list?

Drawing together

● Gather the class together with their lists and compare the different ways children have found to split seven scales between two fish. Compile an ordered list of all the possible answers (there are eight ways). Ask: *Why is it helpful to record our work in this way?*

● Conclude the lesson by referring back to the problem and agreeing that there are eight ways of solving the problem.

Support

Prepare a simple worksheet with eight pairs of fish drawn on it. Ask the children to record their answers pictorially by drawing scales on the fish.

Extension

Challenge the children to split the seven shiny scales between three fish. Ask them to adapt the table and record the answers systematically.

Further ideas

● Set the children similar problems, such as: *How many ways can two children share 12p? Two robots have nine buttons altogether, how many buttons could each robot have?*

● Ask the children to create a similar problem for a friend to solve.

Red packets

Setting the context

Red is a lucky colour to the Chinese. Traditionally red packets, or 'lucky money' envelopes (lei-see), are handed out to younger generations by their parents, grandparents and relatives during Chinese New Year celebrations. A red packet is simply a red envelope with money in it. The envelope is usually decorated with lucky symbols. In China, it is believed that this red envelope will bring luck to the person who gets it and to the person who gives it. The amount of money a child receives in their red packet depends on their age. The older the child is, the more money is likely to be in their red packet.

Problem

This red packet contains two coins. Both of the coins have the same value. How much money could be inside the packet?

Objectives

To solve mathematical problems and puzzles. To double a small number.
To organise the recording of possibilities in an ordered list, starting with the smallest number.

You will need

Two red envelopes; 1p, 2p, 5p, 10p, 20p, 50p, £1 and £2 coins.

Preparation

Put two coins of the same value inside one of the envelopes, and three coins of the same value in the other envelope. Write 'Gung Hei Fat Choy' (Happy New Year) on the front of the envelopes.

Solving the problem

● Share a book or story about the Chinese New Year Festival with the class. Teach the children about the ancient custom of giving red packets during Chinese New Year celebrations. If possible, show the children a picture of a red packet that has been decorated with Chinese writing and lucky symbols.

● Hold up the red packet with two coins inside and read the problem to the children. Ask: *How many coins are inside the packet? What coins could they be?* (Talk about the value of all the coins.) *What do you know about these coins that might help you to solve the problem? What coin would you most/least like to be inside a red packet for you? Why?*

● Ask the children to solve the problem in pairs. Tell them to keep a record of all of the possible amounts of money that could be inside the red packet. Look for examples that show different ways of working to share with the rest of the class during the next part of the lesson.

● Share the children's solutions. Praise children who have found all the possible answers to the problem. Pick different children to explain how they solved the problem and show how they recorded their work.

● Working with the whole class, compile an ordered list of all the possible answers on the board, starting with the smallest amount that could be in the envelope. Model how to record each answer as a number sentence: 1p + 1p = 2p; 2p + 2p = 4p and so on.

● Each time a solution is added to the list, invite the children to say what the next smallest coin in the envelope could be. Encourage the children to see that by working systematically they will know when they have listed all the possible solutions to the problem.

- Choose a child to open the envelope and total up how much money is inside. Refer back to the list on the board to show the children that this was one of the possible solutions they found.

- Show the children the second red packet. Explain that there are three coins of the same value inside this envelope. Tell the children that you would like them to work independently to investigate how much money could be in the packet. Instruct the children to start with the smallest coin that could be in the envelope and then continue recording results in a systematic way to check for all possibilities.

Drawing together

- Share the children's solutions to the second problem. Again, choose a child to open the envelope and work out how much money is inside.

- Choose a child who solved the problem and recorded the results systematically to describe and explain how they organised their work to the rest of the group.

- Encourage the children to notice that the number of possible answers is the same whether there are two or three coins in the red packet. Can anyone explain why this is? Ask: *Would the number of possible answers be the same if there were four coins of the same value? Why?*

Support
Provide hundred-squares or practical counting apparatus to help the children to calculate the value of two coins.

Extension
There are three silver coins in the red packet. How much money could there be altogether?

Further idea
Find out how much money could be in the red packet if there were four coins of the same value inside.

King Lazybones

Setting the context

King Lazybones is a rich and powerful king. He lives in a huge palace and has eight servants who do everything for him: a cook, a cleaner, a gardener, an entertainer, an inventor, a driver, a footman and a knight.

Every time King Lazybones wants something he rings a bell. When the servants hear the bell they have to stop what they are doing and run to kneel before the king. The servants are fed up with rushing to the king's aid only to find out that they are not the servant he needs. The inventor has had an idea – he is going to try and invent eight special codes for King Lazybones to use to summon each of his eight different servants.

Problem

Using two different instruments, invent eight special codes (one for each of the servants). Each code must be made up from three sounds (one or both instruments may be used).

Objectives

To solve mathematical puzzles and problems. To devise a strategy for recording different answers.

You will need

A crown; a throne (drape a piece of material over a large chair); a bell; simple percussion instruments; paper and pencils; individual whiteboards and pens.

Preparation

Pick a child to be King Lazybones and eight children to be the servants.

Solving the problem

● Set the context for the lesson by telling the children about King Lazybones. Help the children to act out the story. Each time King Lazybones rings his bell, tell the servants to stop what they are doing and come to kneel before the king.

● Discuss why the king is called King Lazybones. Why do the children think the servants are fed up? Ask: *How does the inventor's plan overcome the problem?*

● Read the problem with the class and discuss what it is asking them to do. Ask the children to identify the key information given in the problem. List these key points on the board for children to refer back to during the lesson.

● Ask a child to choose two instruments (such as a drum and a triangle) from a selection of simple percussion instruments. Ask the children, in turn, to play three sounds using one or both of the instruments, for example: drum, drum, triangle.

● Organise the children to sit in pairs. Give each pair a small whiteboard and pen. Tell them to listen carefully. Play a three sound pattern. Ask the children to try and record the sounds you played on their boards (they may use words, symbols or pictures).

Invite individuals to hold up their boards and explain how they recorded the sounds. Discuss which of the methods the children think is most efficient. Model how to record the sounds quickly and easily using a simple abbreviation, such as T T D (triangle, triangle, drum).

Ask the children to think of a different three sound pattern that could be played using the same two instruments. Invite them to record it on their boards, using the method you have just modelled.

Let each pair, in turn, play their sound pattern. List all the sound patterns the children play on the board. Check the list for repeats, and cross out any that are the same. Count the number of different patterns in the list. Ask: *Can you think of any different sound patterns we could make with these two instruments?*

Let each pair of children choose two percussion instruments. Ask them to work together to invent eight different three-sound calls (one for each servant). Give each pair of children eight blank cards and a felt-tipped pen. Invite them to record each different sound pattern they create on one of the cards, using the simple recording method you have taught them. Encourage the children to check that each of the sound patterns they record are different.

Drawing together

Gather the children together. Invite a pair of children who have solved the problem correctly to bring their cards to the front and stick them randomly on the board. Ask them to describe what they were trying to find out, and how they know they have answered the problem correctly.

Show the children how to reorder the cards into a systematic list in order to check that no answers have been repeated.

Conclude the story about King Lazybones. Take the cards from the board and give one to each of the king's servants. Retell the story. This time, instead of ringing a bell, ask the king to play one of the three sound patterns. Notice what happens. Only the servant who hears the sound pattern on their card will come to kneel before the king – the rest of the servants can carry on with what they were doing. Problem solved!

Support
Provide the children with simple picture cards to stick down to record different answers.

Extension
Ask: *Using two different instruments, how many different four-sound patterns can you make?*

Further idea
Devise different sound patterns for classroom instructions, such as one for settling down, one for tidying up and so on.

Spaceman Sid

Setting the context

Spaceman Sid travels all over the universe in his spaceship. He has explored many faraway planets and has seen all sorts of things on his journeys through space. Spaceman Sid's latest mission has taken him to a planet called Planet Crit, a small planet inhabited by aliens. The aliens on Planet Crit look very strange. They all have a long antennae sticking out of their head. Attached to these antennae are the aliens' eyes. There are two types of alien: aliens with one eye and aliens with two eyes.

Problem

Spaceman Sid landed on Planet Crit. He met some aliens. Some of the aliens had one eye. Some of the aliens had two eyes. Altogether he counted nine eyes. How many of each type of alien could he have seen? Can you find all the possibilities?

Objectives

To solve mathematical problems and puzzles.
To add more than two small numbers together.
To begin to show awareness of a system for finding all the possibilities.

You will need

Two copies of photocopiable page 29 for each group; one enlarged set of the alien cards from the photocopiable page (coloured and laminated); long strips of paper; Blu-Tack; ten simple alien headbands (see Preparation, below).

Preparation

Make ten headbands using stiff card (five with one eye and five with two eyes).

Solving the problem

● Tell the children about Spaceman Sid and his adventures. Select ten children to come to the front and put on an alien headband. Ask the rest of the class to imagine Spaceman Sid's reaction when he saw all these googly-eyed creatures looking up at him!

● Pick one child to be Spaceman Sid. Ask her to count the total number of alien eyes that she can see. Ask the class to suggest a different way to count the eyes. If necessary, show the children how to sort the aliens into two groups and then count the total number of eyes by counting in twos and then ones.

● Let different children try to use the strategy to count the total number of eyes that different combinations of aliens have.

- Read the problem to the children. Ensure that the children understand what the problem is asking them to find out. Help them to identify the information contained in the problem that will help them to solve it.
- Stick the enlarged set of alien cards on the board. Instruct one of the children to use the cards to find a solution to the problem. Ask the rest of the children to say whether they think the answer is correct and why.
- Remind the children of the counting strategy you modelled earlier in the lesson, by asking questions such as: *Why has Amelia grouped the aliens with two eyes together at the beginning?* Or: *Can you show Lily how she could have sorted the aliens to make it easier to add up the total number of eyes?*
- Organise the children to work in groups of three or four. Give each group some alien cards, a long strip of paper and some Blu-Tack. Ask the children to work together to find a solution to the problem. Tell each group to record their answers by sticking the alien cards onto the strip of paper.
- Gather the children together. Check each group's answer. Are any of the answers the same? Encourage the children to notice that it is much easier to identify repeats if the aliens are ordered in a line with the two-eyed aliens and one-eyed aliens grouped together.
- Invite the children, in their groups, to find all the possible combinations of aliens that Spaceman Sid may have seen. Tell them to record each answer on a separate strip of paper. Remind the children to check their work carefully so that they do not repeat any answers.
- Talk to each group about what they are doing. Show the children how to organise the answers they have found to help them decide whether they have found all the possibilities:

2 + 2 + 2 + 2 + 1
2 + 2 + 2 + 1 + 1 + 1

Ask: *Which strip of aliens comes next?*

Drawing together

- Ask each group to say how many answers they have found. Choose a group who showed good understanding of systematic recording to stick their answers, in order, on

the board. Ask them to explain to the rest of the class how, by organising their answers in this way, they were able to recognise when they had found all the possibilities.
- End the lesson by referring back to the problem and concluding that there are four possible answers.

Support
Provide the number of two-eyed aliens for each combination, asking them to work out the number of one-eyed aliens.

Extension
Ask the children to record each possibility as a number sentence.

Further idea
Ask: *How many different ways can you pay for a toy that costs 8p, using a combination of 1p and 2p coins?*

Boats

Setting the context

Read *The Enormous Crocodile* by Roald Dahl (Puffin Books) or teach the children the rhyme below to introduce the context for this problem-solving lesson.

Row, row, row your boat
Gently down the stream
If you see a crocodile
Don't forget to scream!

Problem

Twelve children want to cross a stream to meet their friends on the other side. There are three boats tied to the bank. Next to the boats there is a sign that reads: 'BEWARE OF THE CROCODILES!'.

If the children wish to cross the stream safely they must row all three boats across the river at the same time; there must be an even number of passengers in each boat. Ask: *How many children could go in each boat? Can you find different ways to do this?*

Objectives

To solve simple problems and puzzles.
To recognise odd and even numbers.

You will need

Eight large mats (boats); green beanbags (crocodiles); a large space; a 'Beware of the crocodiles!' sign, drawn onto a large piece of card; photocopiable page 30; 12 counters/cubes and three pots for each group.

Preparation

Spread the mats around the hall, scatter the green beanbags around the mats.

Solving the problem

● Ask the children to imagine that the mats are rowing boats floating on a crocodile-infested stream. Instruct all of the children to get into one of the boats.

● Play 'Boats'. Explain that you would like the children to skip up and down the hall until they hear you shout, 'Even!'. Explain that they must then get into the boats as quickly as they can, ensuring that there is an even number of people in each boat. After 20 seconds, any children who are not in a boat or who are in a boat with an odd number of children will be eaten by the crocodiles! Play

the game several times, varying the number of boats by adding or taking away mats.

● Place three mats (boats) at one side of the hall. Display the 'Beware of the Crocodiles' sign beside the boats. Introduce the problem.

● Pose the following questions to check that the children have understood the problem: *How many children want to cross the stream? How many boats are there? What do you know about the number of children that can cross the stream in each boat?*

● Choose 12 children to act out the problem. Invite them to organise themselves into the three boats so that they can cross the river safely. Count how many children are in each boat. Record the information on an enlarged copy of photocopiable page 30. Ask the rest of the class to say whether this is a correct solution to the problem.

● Back in the classroom, organise the children to solve the problem in groups of three. Provide each group with a copy of photocopiable page 30, 12 counters/cubes (children) and three pots (boats).

● Ask the children to find as many different solutions to the problem as they can and to record the solutions on the photocopiable sheet. (If necessary, list all of the even numbers to 12 on the board for reference.)

● As the children are working, talk to them about what they are doing. Encourage them to check their work to make sure that they do not repeat any answers. Develop the children's understanding of the problem-solving process by asking questions such as: *How did you get this answer? What is the largest number of children that can go in a boat? Why?*

Drawing together

● Gather the children together. Compare the number of different solutions each group has found to the problem. Do any of the groups think they have found all of the possible answers? How can they be sure? What can they do to check their solution?

● Demonstrate how to compile a systematically-ordered list of all the possible answers. Ask 12 children to stand by the three boats. Ask: *What is the biggest even number of children that could go in the first boat?* (Establish that it is 8 – it cannot be 12 or 10 because that does not leave enough children to make an even number in the remaining two boats.) *What is the next largest number?*

8 2 2
6 4 2
6 2 4
4 6 2
4 4 4
4 2 6
2 8 2
2 6 4
2 4 6
2 2 8

● Agree that there are ten different ways that the children could cross the river. Explain that by recording the possibilities in an ordered list, starting with the largest

number, you can be sure that you have found all the possibilities.

Support
Simplify the problem by saying: '12 children need to cross the stream in three boats. How many children could go in each boat?'

Extension
Challenge the children to find out if there would be the same number of possibilities if an odd number of children had to travel in each boat.

Further idea
Set the same problem in a different context, for example, Easter eggs in baskets, or cakes on plates.

Quicksand

Setting the context
Introduce the pirate theme by reading a story about pirates and showing the children a simple pirate treasure map.

Problem
The pirate is standing on stepping stone number 5. The treasure is on stepping stone number 12. How can the pirate reach the treasure in two rolls of a dice? Find all the different ways the pirate can do this.

Objectives
To solve mathematical problems and puzzles.
To count on from any small number.
To begin to have a system for finding all the possibilities.

You will need
A book about pirates; a simple pirate costume such as an eye patch or headscarf; 16 small stepping stones cut from thin card numbered 1-12; a bag of treasure (a bag filled with small tokens such as stickers); dice; a copy of photocopiable page 31 for each child in the class.

Preparation
Arrange the stepping stones in order; dress one of the children up in the simple pirate costume.

Solving the problem
● Ask the children to sit in a circle around the stepping stones. Instruct the pirate to stand on the first stepping stone (1) and place the treasure on stepping stone number 8. Tell the children that the pirate is trying to reach the treasure, but he must stay on the stepping stones as they are surrounded on all sides by dangerous quicksand. Make sure that the children know what quicksand is and that they understand what would happen to the pirate if he stepped onto the quicksand!

● Explain to the children that the pirate must get to the treasure in two moves. Roll the dice and instruct the pirate to take that number of steps forward on the stepping

stones. Roll the dice a second time and tell the pirate to count on that many more steps. If the pirate lands on the eighth stepping stone they may take a piece of treasure out of the bag.

● Let different children have a turn to be the pirate. Each time a pirate lands on the treasure, let her take a piece of treasure out of the bag and model how to record the two moves the pirate made as a number sentence. Encourage the children to notice that there is a variety of different ways that the pirate can reach the treasure in two moves.

● Once the children are confident with the game, ask them to work out, by counting on, what number they need to roll with the second dice in order for the pirate to land on the treasure.

● Ask the children to look carefully at the number sentences on the board. Ask: *Are any of the number sentences the same? Can you think of any other ways the pirate could get to the treasure in two moves? Do you think we have listed all the possible answers?*

● Show the children how to order the number sentences systematically by starting with the largest number. Explain that this way it is possible to track whether all the possible answers have been found.

● Give each child a dice. Read the problem to the class. Illustrate the problem by asking the pirate to stand on number 5 and moving the treasure onto stepping stone number 12.

● Invite the children to describe the problem in their own words. Ask: *What is the*

biggest number of steps that the pirate could take on her first move? Show me on your dice (6). Count on to find out how many more steps the pirate needs to take to reach the treasure. Show me the answer with your dice (1). Write 6 + 1 on the board.

● Give each child a copy of photocopiable page 31. Ask the children to find all the different ways that the pirate can get from 5 to 12 in two moves. Encourage the children to try and record their answers systematically.

Drawing together

● Ask the children to say how many different answers they have found. Work together to compile a systematic list of the children's answers. Ask: *What is the biggest number we could have rolled with the first dice? What would we need to roll with the second dice to make 12? What would the next biggest number have been?*

● Continue to list the answers systematically until all the possibilities have been found. Count how many different ways the pirate can get to the treasure in two moves. (The answer is six.) Ask the children

to say why it is useful to record the answers in this way.

Support

Establish that to move from the fifth to the twelfth stepping stone the pirate needs to take seven steps. Provide the children with seven cubes. Ask them to find all the different ways of splitting seven cubes, recording their answers pictorially or in a number sentence.

Extension

Ask: *How many ways can the pirate get to the treasure in three moves?*

Further idea

Set a similar problem in a different context, such as a fireman taking steps up or down a ladder.

Page

19

A walk in the woods

Setting the context

Once upon a time a little girl called Red Riding Hood lived with her mother in a cottage at the edge of a deep, dark wood. Red Riding Hood's grandmother lived in a cottage at the other side of the wood. In the middle of the wood lived a big, bad wolf! One sunny morning Red Riding Hood's mother said, 'Your grandmother is not feeling very well today. I would like you to take this basket of freshly baked cakes over to her cottage. Don't talk to strangers on the way and keep away from the big, bad wolf's den.'

Problem

Can you help Little Red Riding Hood find a safe route through the wood to Grandma's house? How many different routes can you find?

Objectives

To solve mathematical problems or puzzles. To describe positions and directions.

You will need

A simple prop for each character in the story, such as a basket for Red Riding Hood, glasses for grandmother and a knife and fork for the wolf. A copy of photocopiable page 32 and a counter for each child/pair of children.

Preparation

Mark out a 3x4 grid on the floor using chalk, masking tape or 12 blank floor tiles (see photocopiable page 32). Display the problem on the board.

Solving the problem

● Pick three children to be the characters in the story. Give them each the relevant prop and direct them to stand in the correct square on the 3x4 grid (see page 32).

● Tell Little Red Riding Hood to take the basket to Grandmother's house. Explain that she must step from square to square on the floor grid (not diagonally) without passing through the square occupied by the wolf.

● Let other children in the class have experience of walking a route through the wood. Ask each child to describe the route they take using appropriate mathematical language (left, right, forwards).

● Make sure that the children understand that there are lots of different routes that Red Riding Hood could take through the wood. Read through the problem with the class. Explain that you would like the children to try and find all the different routes that Red Riding Hood can take to get to Grandma's house without going through the wolf's den.

● Ask the children to suggest how they could record each of the different routes they find. Demonstrate how the routes could be recorded in a list such as: Forward 3, Right, Forward 4. Alternatively each different route could be drawn onto one of the grids on the photocopiable page.

● Organise the children to work in pairs. Give each pair of children a copy of photocopiable page 32 and a counter. Re-read the problem with the class. Explain that you would like the children to find and record as many different routes as they can using one of the recording strategies you have shown them. Tell the children to keep checking their work to ensure that each of the routes they record are different.

Drawing together

● Ask different pairs of children to come to the front and describe one of the routes they have found. Instruct the rest of the class to check the route being described by moving their counter on their grid.

● Ask: *How many different routes did you find? Which was the longest/shortest route? Do you think you have found all the different routes? How do you know?*

● Demonstrate how to list all the possible routes in a systematic way (Red Riding Hood takes seven routes). Establish that when Red Riding Hood leaves her cottage she can either go forwards or turn right. Begin by finding and listing all the routes that begin by going forwards and then list all the routes that begin with a right turn.

● Make sure that the children understand that by recording the routes systematically they are able to ensure that they do not miss out any solutions and that they do not record the same route twice.

Support

Give the children more opportunities to actually walk through different routes on the large grid. Ask the children to find and record four different routes that Red Riding Hood could take.

Extension

Adapt the grid by making it larger to allow for more possible routes. Ask the children to predict and investigate whether there would be more or less safe routes to Grandma's house if there were two big bad wolves in the wood or if the wolf's den was located in a different part of the wood.

Further ideas

● Ask the children to create a similar problem by devising a grid based on a different story. Pick one of the children's problems for the class to solve in a subsequent lesson.

● Design a simple 3x4 grid. Ask the children to investigate how many ways a rocket can travel to the moon without crashing into an alien spaceship.

Summer holiday

Setting the context
Introduce the problem using a teddy bear to engage and motivate the children. Choose a name for the teddy and explain that he is going on holiday to Spain. Show the children a picture of his destination (from a travel brochure or postcard).

Problem
Teddy is going on holiday to Spain. He has packed four pairs of shorts and three T-shirts in his suitcase so that he has just enough clothes to be able to wear a different outfit on each day of his holiday. Can you work out how many days Teddy is going on holiday for?

Objectives
To solve mathematical problems and puzzles. To begin to learn a system for finding all the possibilities.

You will need
A teddy; a picture of a warm holiday destination, such as Spain; a world map or globe; a copy of photocopiable page 33 for each child; thin card (red, blue, yellow, black, green); coloured crayons; suitcase containing holiday items (optional).

Preparation
Cut out seven simple cardboard shapes to represent the clothes in Teddy's suitcase (three T-shirts and four shorts in the colours described on photocopiable page 33).

Solving the problem
● Explain that Teddy is going on holiday to Spain. Help the children to locate Spain on a world map or globe and show them a picture of Spain in a travel brochure or on a postcard. If any of the children have been to Spain, ask them to describe their experiences: *What did they do there? Where did they stay? What was the weather like?*
● Ask the children to say what things they

think Teddy will need to pack in his suitcase for his holiday. (You could pack a selection of items in Teddy's suitcase and ask the children to guess what is inside.)
● Display the problem on the board and read it to the children. Ask the children to talk about the problem with a partner. Find out what they have understood by asking questions such as: *What do you need to find out? What do you already know that might help you to solve the problem? What equipment/strategies could you use to help you?*
● Discuss the children's ideas. Make sure that all the children understand that to work out the answer to the problem they need to explore how many different outfits can be made by putting the T-shirts and shorts together in all the possible combinations.
● Stick the cardboard shorts and T-shirts on the board. Let the children take it in turns to make an outfit for Teddy. Establish that there are many different outfits that Teddy could wear and in order to be able to solve the problem the children will need to record each of the different outfits they make. Ask the children to suggest how they could record this information (for example, by drawing, listing or mapping).

Give each child a copy of photocopiable page 33. Explain that you would like them to record all the different outfits they can make using the clothes in Teddy's suitcase by colouring in the pictures on the sheet.

Ask the children to swap sheets with the person sitting next to them. Tell them to compare their answers and then to check each other's work for repeats. Hopefully, there will be some variation in the children's answers to use as a teaching point in a plenary session!

Drawing together

List the children's answers on the board. Tell the children that you are going to solve the problem yourself.

Explain that when you are solving a problem such as this one, you find it helpful to use a system for ordering your work. In simple terms, describe and model how by keeping one variable of the problem fixed while changing another you can keep track of which outfits you have already recorded and which you still have left to find.

Start by working out how many times Teddy can wear his red shorts. Record all the possibilities on an enlarged copy of photocopiable page 33. Find and record all the different outfits that can be made with the black shorts and finally, with the blue shorts.

Ask: *Do you think there are any more outfits? How do you know?* Make sure that the children understand the usefulness of taking a systematic approach to solving and recording a problem such as this one.

Count all the possibilities. Conclude that you can make 12 different outfits from the clothes in Teddy's suitcase. The answer to the problem is therefore, that Teddy is going on holiday for 12 days.

Support
Give the children their own set of cardboard shorts and T-shirts. Tell them to make each different outfit before recording it on the photocopiable sheet.

Extension
Teddy adds two pairs of shoes to his suitcase. How many different outfits can he make now?

Further idea
Using three different coloured scarves and three different coloured hats, how many different outfits can you make for a snowman?

Page
23

Colourful caterpillars

Setting the context

Introduce the caterpillar theme by reading *The Very Hungry Caterpillar* by Eric Carle (Puffin). Show the children an assortment of caterpillar pictures. Spend a few minutes discussing the pictures with the class. Ask the children to describe the caterpillars. Focus primarily on the colours and patterns that distinguish each of the different types of caterpillar.

If possible teach this lesson as a follow up to 'Summer Holiday' (pages 22-23). This will give you an opportunity to see which children (if any) use a system for finding and recording all the possibilities such as the one you demonstrated during the plenary of the previous lesson.

Problem

How many different caterpillars can you make using four different colours? Each section of the caterpillar must be a different colour.

Objectives

To solve mathematical puzzles and problems.
To check for repeats of possibilities.
To begin to have a system for finding all possibilities.

You will need

Four large circles of paper (red, yellow, blue, green); one caterpillar head and tail; counters; small squares of paper; paint and corks or small circles of paper and glue; pictures of caterpillars with distinctive patterns.

Preparation

Put trays of paint (red, blue, yellow, green) and four corks on each table.

Solving the problem

● Introduce the context for this activity by reading the story and looking at the collection of caterpillar pictures (see above).
● Read the problem to the children and establish what the problem is asking them to find out. Ask: *What rules must you follow? What equipment could you use to help you?*

● Make a caterpillar by sticking the four large paper circles in a row on the board. Pick a child to come and rearrange the circles to make a different caterpillar. Repeat this process several times. Establish that there is a variety of different caterpillars that can be made using four colours. Remind the children that the problem asks how many different caterpillars can be made using four colours. Ask the children to predict how many possible outcomes there might be.
● Divide the class into small groups. Send each group to sit at a different table. Explain that you would like the children to work together to design and record as many different caterpillars as they can using four colours.
● Give each group four coloured counters (red, blue, yellow, green). Before the children begin, show them how to design a caterpillar by arranging the counters in a line. Then, demonstrate how to use the paint and corks to print a record of the caterpillar. Tell the children to print each caterpillar on a different piece of paper.

● Observe and interact with the children as they work. Encourage the children to discuss their work and interpret the information they have gathered in relation to the problem. Ask: *How many different caterpillars have you found so far? How do you know you have you followed all the rules of the problem?*

● Remind the children to check their work carefully for any repeats. As the caterpillars are printed on separate pieces of paper they can identify and remove any duplicates.

● After about ten minutes tell the children to stop what they are doing. Ask each group to say how many different caterpillar designs they have made so far. Instruct the children to organise their caterpillars into sets, according to their start colour. Explain that by organising the caterpillars in this way it will be easier to see if there are any designs still to make.

© Photodisc Inc.

Drawing together

● Gather the children together. Ask each group, in turn, to say how many different caterpillars they have made and whether or not they think they have found all the possibilities. Blu-Tack one group's printed caterpillars on the board. Organise the caterpillars into four lists according to their start colour.

● Assess and develop the children's understanding of the problem by asking the following questions: *What do you notice about the number of caterpillars in each list? Why do you think there are the same number of caterpillars in each list?* (If there are not the same number in each list you will need to ask questions to establish whether or not the children realise that there should be.)

● Now ask: *Do you think this group have found all the caterpillars it is possible to make with four colours? How do you know?* (The maximum number of possibilities is 24.)

● Finally, ask the children to say whether they found it easier to solve the problem and check work before or after they had organised the caterpillars into lists. Why?

Support
Ask the children to explore how many different caterpillars they can make with three colours.

Extension
Investigate how many different caterpillars you can make using five different colours.

Further idea
How many different flags can you make using four colours? You must use each colour every time.

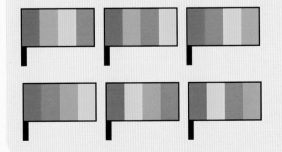

Joseph's coat of many colours

Setting the context

The story of Joseph and his coat of many colours provides a creative context for this problem solving activity. If possible teach this lesson as a follow up to 'Colourful caterpillars' (pages 24-25). This will give you an opportunity to observe which children are able to independently implement the system for finding all the possibilities that you modelled at the end of that lesson.

Problem

Using eight squares of material (two yellow, two green, two red, two blue) design Joseph a symmetrical coat of many colours. How many different symmetrical coats can you make?

Objectives

To solve mathematical problems and puzzles.
To recognise line symmetry.
To work systematically and explain methods and reasoning.

You will need

The story of Joseph's coat from a children's bible or similar; a copy of photocopiable page 34 for every child; an enlarged, laminated copy of Joseph's coat (page 34); Multilink cubes; crayons; paper (yellow, green, red, blue); mirrors.

Preparation

Cut the coloured paper into small squares that are the same size as the patches on the laminated copy of Joseph's coat (two of each colour).

Solving the problem

● Tell the story of Joseph and his coat. Show the children a picture of Joseph in his coat. Discuss the simple T-shaped tunic coats worn in biblical times. Ask the children to describe the colours and patterns on Joseph's coat.

● Stick the laminated template of Joseph's coat onto the board. Read the problem to the children. Ask the children to describe the problem in their own words. Check that the children understand the meaning of the term symmetrical and ask them to identify the line of symmetry on the coat template.

● Invite one of the children to stick the eight small paper squares onto the coat to make a symmetrical pattern. Discuss the design with the class. Ask: *Is the pattern on the coat symmetrical? How do you know? What could we use to check the design is symmetrical?* Demonstrate how to use a mirror to check the coat design is symmetrical.

● Swap two of the squares over on one side of the coat. Ask: *Is the coat symmetrical now? Why not?* Let one of the children come and adjust the position of the squares on the other side of the coat to correct the design.

● Give each child in the class a copy of photocopiable page 34. Instruct the children to design a symmetrical coat by placing cubes/small squares of paper onto the coat template. Show the children how to make a record of the design by colouring in one of the smaller coats.

Remind the children that in 'Colourful caterpillars' (see pages 24-25) you showed them a method of solving the problem systematically to make sure that you found all the possible solutions. Can the children remember what this was? (Finding and recording all the caterpillars beginning with a red circle, then all the caterpillars beginning with each of the other colours in turn.)

Discuss how the children could apply the same strategy to this problem. For example, finding all symmetrical coats with a yellow sleeve, then all the coats with a blue sleeve and so on.

Observe how the children approach the task. Notice which children use a 'hit and miss' approach to finding all the possible coat designs and make sure you remind these children to check carefully for repeats before recording each possibility. Look for examples of children who employ a system for finding all the possibilities.

Drawing together

Gather the class together. Pose the following questions: *What did the problem ask us to find out? What equipment did you use to help you solve the problem? How did you record your work?*

Ask one of the children who used a systematic approach to describe their strategy to the rest of the class. Ask the child to explain how they knew when they had found all the possible outcomes.

Illustrate this systematic way of working by working through the problem with the class. Start by finding all the coats with a yellow sleeve. Record them quickly and simply as follows:

Ask the children to predict whether or not the

number of coats with blue, red and green sleeves will be same and why.

Conclude the lesson by referring back to the problem and agreeing that there are 24 different symmetrical coats that can be made using the eight coloured patches. Ask the children to say what they might do differently if they were solving the problem again.

Support
Reduce the number of possibilities by blanking out the bottom two squares on the coat template and giving the children six patches instead of eight.

Extension
Challenge these children to record their work systematically.

Further idea
Challenge the children to find out how many different ways they can use three colours to colour in the spots on a butterfly's wings. The pattern must be symmetrical every time.

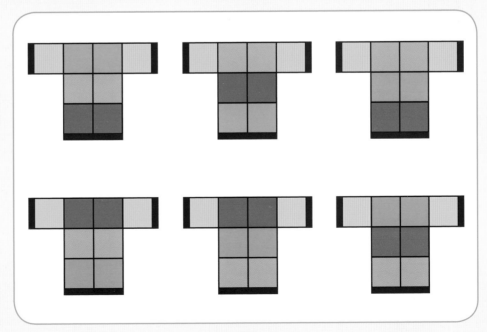

The Rainbow Fish

- How many different ways can the Rainbow Fish share his scales with his friend the Little Blue Fish?

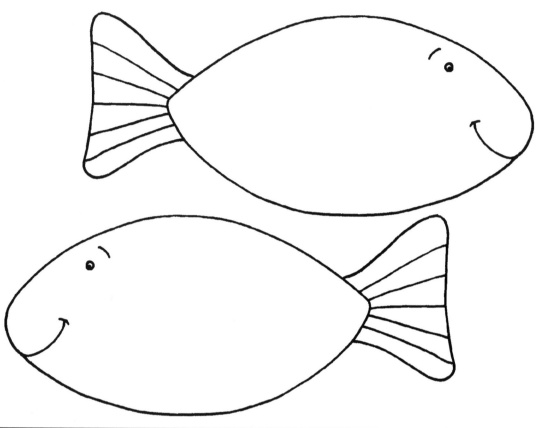

Rainbow Fish	Little Blue Fish

Creative Activities for Maths Problem Solving: Ages 5-7

■SCHOLASTIC
www.scholastic.co.uk

Spaceman Sid

Boats

- 12 children want to cross a stream to meet their friends on the other side. There are 3 boats tied to the bank. Next to the boats is a sign that reads:

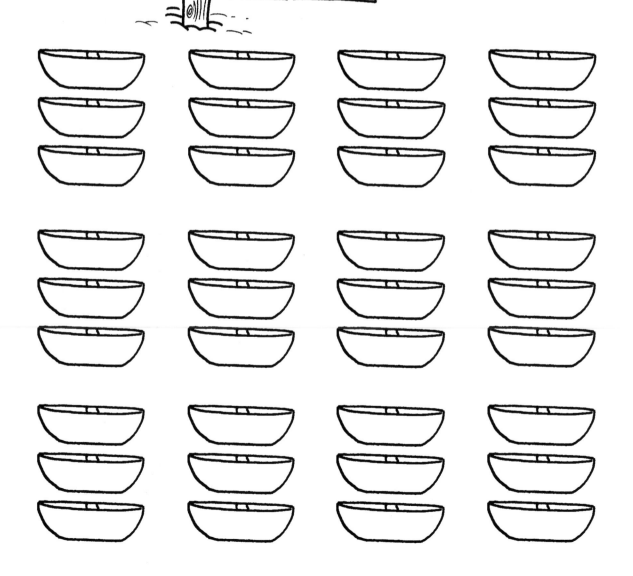

BEWARE OF THE CROCODILES!
If you wish to cross the stream safely you must row all three boats across the river at the same time.
There must be an even number of passengers in each boat.

- How many children could go in each boat?

- How many different ways can you find to do this?

Quicksand

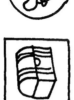
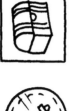

- The pirate is standing on stepping stone number 5.

- The treasure is on stepping stone number 12.

- How can the pirate reach the treasure in two moves?

- Find all the different ways the pirate can do this.

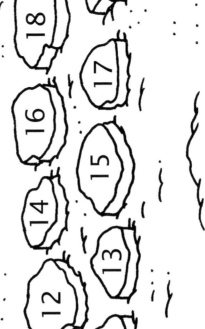

PHOTOCOPIABLE **Creative Activities for Maths Problem Solving: Ages 5-7**

A walk in the woods

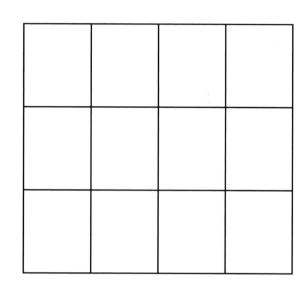

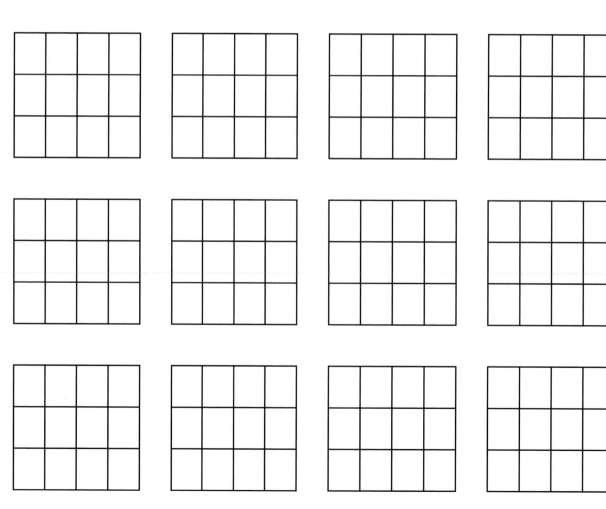

■SCHOLASTIC
www.scholastic.co.uk

Summer holiday

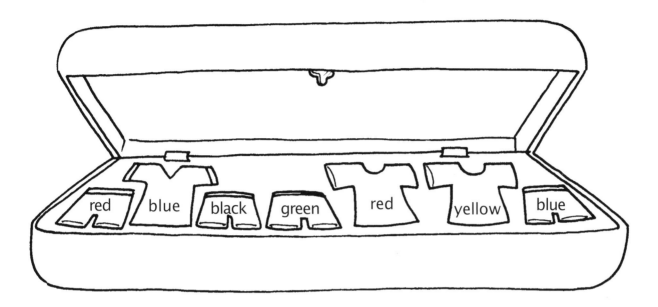

red   blue   black   green   red   yellow   blue

Joseph's coat of many colours

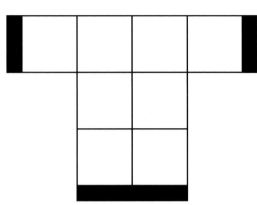

● Use eight squares of material (two yellow, two green, two red and two blue) to design Joseph a symmetrical coat of many colours. How many different symmetrical coats can you make?

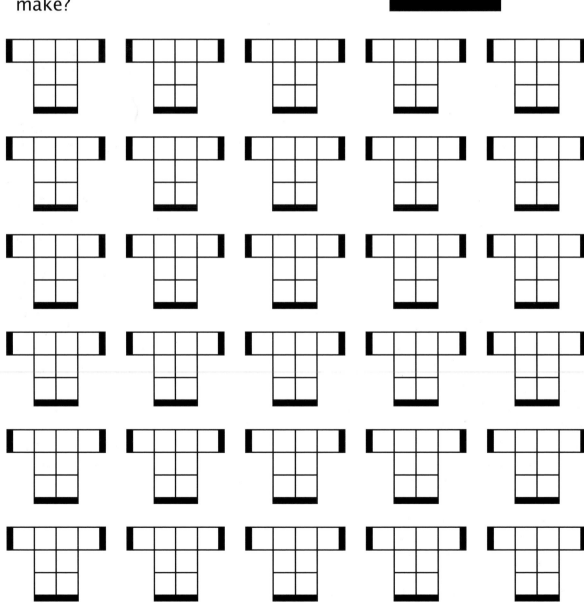

■SCHOLASTIC
www.scholastic.co.uk

Chapter Two

Finding rules and describing patterns

The activities in this chapter are specifically designed to help the children to develop the skills and strategies they need in order to tackle problems that involve finding rules and describing patterns. To solve this type of problem it is necessary first for the children to identify, describe and then continue a pattern, before they then determine a rule for the pattern that can be used to predict other items in the sequence.

The first three problems in this chapter are based on simple repeating patterns. The children are shown how to record and continue patterns using very practical, visual methods – for example, acting out the problem ('Take the train', see pages 36-37) or drawing ('Sidney snake', see pages 40-41).

As the chapter progresses, the problem-solving activities gradually increase in complexity. The children are taught to identify and describe patterns in number sequences and devise simple rules for the number sequences generated by different patterns.

Children should be taught a range of appropriate skills and strategies that they can use to help them solve these types of problems. These include how to:

- read the problem carefully and identify the question;
- identify the key information contained in the problem;
- continue a pattern by acting the problem out, through using equipment, drawing or listing a number sequence;
- record patterns clearly; for example, in a list or table;
- make up a rule for a pattern and use the rule to predict the next few terms in the sequence.

Take the train

Setting the context
Read the story *Teddybears Take the Train* by Susanna Gretz (A & C Black) as an ideal introduction to this lesson.

Problem
Nine teddies are in the queue for an empty train. Which carriage will the ninth teddy be in?

● Three teddies can sit in each carriage of the train.
● The carriages must fill up from the front first.
● A teddy cannot sit in the next carriage unless all the ones before it are full.

Objectives
To solve mathematical problems and puzzles. To describe and continue a simple pattern. To devise and use an appropriate problem-solving strategy.

You will need
Teddybears Take the Train by Susanna Gretz (A & C Black); several small mats (carriages); paper and pencils; Compare Bears; a set of bear cards for each child (photocopiable page 52); labels: '1st', '2nd', '3rd', and so on.

Preparation
Make a simple train by arranging the mats (carriages) in a line, leaving a small gap between each one. Place a chair at the front of the row of mats for the driver to sit on.

Solving the problem
● Choose a child to be the train driver. Tell her to sit on the chair. Ask the rest of the class to sit in a 'carriage'. Read the story book to the class.
● Briefly revise the children's knowledge of ordinal numbers. Instruct the children to stand up if they think they are in the third carriage, fifth carriage and so on. Label each of the carriages together – 1st, 2nd, 3rd and so on.
● Introduce the problem. Ask: *What do we need to find out? What do we know that might help us to solve the problem? What could we use to help us solve the problem?*
● Ask nine children to line up. Ask the children to imagine that these children are the nine teddies waiting for the train. Tell the teddies that when the train arrives, they must get onto the train in an orderly fashion. Starting from the front of the train, they must fill each carriage in turn.
● Act out the problem. Stress how important it is that the children follow the rules of the problem. Which carriage is the ninth teddy in?
● Act out some more problems, using different numbers of 'teddies' each time. Use the children and then Compare Bears to act out each scenario.
● Give the children, in pairs, a small whiteboard and pen. Challenge them to predict which carriage the specified teddy will be in each time. Observe the range of strategies children use to predict the answer. Do they guess, try to draw the problem, count in threes...? Act out the problem to work out the answer each time. Did anyone predict the answer correctly? Ask them to explain the strategy they used to the rest of the group.
● Change the number of teddies to 24. Challenge the children to solve the problem by themselves. Explain that you would like the children to devise their own strategy for solving the problem using their whiteboard and pen or a piece of paper and a pencil.

● As the children work, question them to make them think about the strategy they are using to solve the problem. Ask: *Describe how you are working out the answer. How are you keeping track of the number of teddies in the train so far?* Encourage these children to look for patterns in the data. Can they solve the problem by counting in threes?

Drawing together

● Let different children describe the strategy that they used to solve the problem. What information did they use to help them describe and continue the pattern?

● Demonstrate how to record the pattern quickly and concisely in the form of a list:

Carriage 1	1, 2, 3
Carriage 2	4, 5, 6
Carriage 3	7, 8, 9

● Encourage more able children to notice that if the number of bears is a multiple of three, it is easy to calculate which carriage the last bear will be in by counting in threes.

● Assess which children understand and can apply this rule by asking the children to count in threes to work out which carriage the 12th, 15th or 18th bear would be in.

Support

Demonstrate how to solve the problem practically by counting out the correct number of Compare Bears and then organising them into groups of three. Provide a set of bear cards for each child so that they can solve the problem practically.

Extension

Ask the children to investigate a more complex pattern by including the driver in the pattern. The pattern will then become 1 + 3 + 3. Challenge the children to use this rule to predict which carriage the 20th bear would be in.

Further idea

Set the children a similar problem in a fairground context. For example, 17 children are queuing up for the roller coaster. Four children can go in each roller coaster carriage. Which carriage would the 17th child be in?

Marching

Setting the context

The traditional nursery rhyme, 'The Grand Old Duke of York', provides the context for this problem-solving lesson. Recite the nursery rhyme together. Practise marching in time like real soldiers and saying, 'Left, right, left, right...'

Problem

The Grand Old Duke of York marched his men to the top of the hill. The soldiers marched left, right, left, right for a total of 30 steps. Were the soldiers marching on the left or the right side when they took their 30th step?

Objectives

To solve mathematical problems and puzzles. To describe and continue a repeating pattern. To recognise odd and even numbers.

You will need

Paper and pencils; space for the children to march.

Solving the problem

● Start a repeating pattern using simple actions that the children can copy, such as 'tap, clap, clap' and so on. Ask the children to join in and follow the pattern. Create further patterns for the children to join in with. Gradually increase the complexity of the patterns as the children gain confidence.
● Sit in a circle. Start a pattern. Ask the children to take it in turns around the circle to add the next action. This will give you an opportunity to assess which children are able to continue a pattern. Target specific children, ask them to describe the pattern and explain how they knew which action came next. Let individual children invent patterns for the rest of the class to follow.
● Sing the 'Grand Old Duke of York'. Let the children pretend to be soldiers and accompany the nursery rhyme with appropriate actions. Choose one child to be the Grand Old Duke of York and march at the front of the group. Ask him to lead the rest of the class, marching in time and saying, 'Left, right, left, right....'. This may take a bit of practice!
● Give the children, in pairs, a piece of paper and a pencil. Ask them to try and think of a way to record the marching pattern. They may decide to draw pictures, write words or letters to represent the pattern. Discuss and evaluate the children's ideas. Ask them to decide which of the recording methods they think is the most efficient.

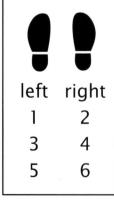

left   right

1	2
3	4
5	6

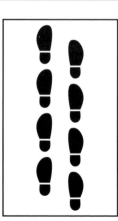

1. L

2. R

3. L

4. R

5. L

● Introduce the problem by reading it to the children. Ensure that the children understand that in order to solve the problem they need to continue the pattern. State that you would like them to keep a record of what they do by using one of the recording strategies discussed earlier.

● While the children are working, talk to them about the strategies they are using to record the soldiers' marching pattern. Notice which children stop to count up how many steps they have recorded so far. Ask questions which will guide these children towards writing the numbers 1 to 30 alongside the pattern of steps – a much more efficient strategy for keeping track of how many steps they have recorded.

Drawing together

● Discuss the children's solutions to the problem. Select a few children to talk about how they solved the problem and to show how they recorded the pattern. Ask the rest of the class to say what they think is good or could be improved about the recording methods the children have used.

● Demonstrate to the children how you would have solved the problem by recording the pattern in a list. (If someone in the group has done this, you could simply use their work as an example.)

L   R
1   2
3   4
5   6 ...

● Ask the children to look carefully to see if they notice anything about the pattern of numbers in the table. Hopefully, someone in the group will notice the rule that on odd numbers soldiers march to the left; and on even numbers, they march to the right.

● Ask the children to use the rule to say which side the soldiers would march on by their 43rd stride.

Support
Simplify the problem by asking the children to say whether the soldiers were marching on the left or right when they took their 12th step. Show the children a strategy for recording the marching pattern visually. Draw a row of 12 footprints and annotate each footprint L or R.

Extension
Ask the children to describe the rule of the sequential pattern in words. Ask: *What would the rule be if the soldiers' first step was on their right foot?*

Further idea
Set a 'Cheerleaders' problem – devise a simple routine such as 'jump, clap, wave'. Ask the children to continue the routine. What will the 20th action be?

Sidney snake

Setting the context
Sidney snake is a colourful, stripy snake. He has 25 stripes altogether – some red, some yellow and some blue. Sidney's stripes follow a special repeating pattern: red, red, blue, blue, yellow.

Problem
What colour is Sidney snake's last stripe? How many yellow stripes does Sidney snake have altogether?

Objectives
To solve mathematical problems and puzzles. To describe and continue a repeating pattern. To recognise multiples of 5.

You will need
A selection of equipment that children can use to represent the pattern of Sidney snake's stripes (such as red, blue and yellow beads, cubes, counters, finger paints and crayons).

Solving the problem
● Start a repeating pattern using the children in the class (for example, long hair, short hair, long hair). Tell the children that you would like them to help you to continue the pattern. Let different children predict the next few terms in the sequence. Ask them to suggest who could come and join the line next. Why? Invite one of the children to describe the rule of the pattern.

● Make up other repeating patterns using different attributes of the children, for example, eye colour or hair colour.

● Tell the children about Sidney snake. Read the problem to the class and ask the children to identify the two things that the problem requires them to find out – the colour of Sidney snake's 25th stripe and the total number of yellow stripes that are on Sidney snake's body.

● Ask the children to think about how they might solve the problem. Make sure that all the children understand that they need to find a way of representing the 25 stripes on Sidney snake's body. Ask them to think about

the sort of equipment they could use to do this, for example, coloured cubes, beads, counters, crayons or paints.

● Organise the class to solve the problem in pairs. Where possible, let the children make their own decisions about the equipment they would like to use to illustrate the pattern of stripes on Sidney's body.

● As the children work, ask them to respond to questions that will focus their attention on the strategy and patterns they are using to solve the problem. Ask: *What are you doing? What colour will the next stripe in the pattern be? How do you know? What do we call this type of pattern? How are you keeping track of the number of stripes in the pattern so far? How will you know when you have reached the 25th stripe?*

● Look out for children who are writing the numbers 1-25 alongside the pattern to help them to keep track of how many stripes they have recorded. Show their work in progress to the rest of the group and recommend that everybody uses this strategy to help them keep track of when they reach the 25th stripe.

Drawing together
● Share the children's solutions to the problem. Does everyone agree that the 25th stripe is yellow and that Sidney snake has five yellow stripes altogether? Develop the children's reasoning and communication skills. Ask them to describe how they worked out the answer to the problem and why they decided to organise their results in a particular way.

- Ask the children to consider how they could have solved the problem and organised the results if you had only allowed them to use a paper and pencil.

Show the children how they could have recorded the information in a list.

- Highlight the position of the five yellow stripes in the sequence. Ask the children to say what they notice about the position of the yellow stripes. (They are all a multiple of 5.) Ask the children to imagine that Sidney snake had a longer body. Ask: *What would the position of his next yellow stripe be? How many stripes long would he have to be to have 10 yellow stripes?*

Support
Ask less-able children to repeat a simpler repeating pattern, such as: red, yellow, red...

Extension
Tell the children that Sidney's dad has the same stripy pattern as Sidney, but he is much longer. Ask: *What colour would his 40th stripe be? How did you work it out?*

Further idea
Ask the children to make up a similar problem in a different context, such as stripes on a scarf, flowers in the garden or front doors along a street.

Teddy bears' picnic

Setting the context
It was a lovely summer's day. The Three Bears decided that it was too hot to have porridge for their breakfast. Instead Father Bear suggested that he make one of his delicious giant honey sandwiches and they take it into the woods to have a picnic breakfast.

Problem
How many cuts would you have to make so that each bear can have a piece of sandwich? It does not matter if the pieces of sandwich are different shapes or sizes, but you have to cut each piece individually.

Objectives
To solve mathematical problems and puzzles.
To identify, describe and continue patterns in a sequence of numbers.
To use a rule to predict the next few numbers in a sequence.

You will need
A collection of teddy bears; a picnic blanket; plates; cups; a knife and a giant honey sandwich (made from card or sponge); scissors; lots of small square pieces of paper (sandwiches); photocopiable page 53.

Preparation
Spread the picnic blanket on the floor. Sit three teddies on the rug and give each teddy a plate and cup. Put the giant honey sandwich on a plate in the middle of the rug.

Solving the problem
● Organise the class so that they are sitting around the edge of the picnic rug. Introduce the context for the problem-solving activity using the simple scenario above.
● Read and discuss the problem with the class. Pose some simple questions to make sure that the children understand what the problem requires them to do. Ask: *What is*

the problem asking us to find out? How many pieces do we need to cut the sandwich into so that each bear can have a piece? What equipment could we use to help us act out and solve the problem?

● Divide the class into groups of three. Give each group a piece of paper, a pencil and a pair of scissors. Ask the children to act out the problem, pretending that they are the three bears. Invite them to cut up the sandwich (piece of paper) so that they can each have a piece. Tell the children to count how many cuts they need to make to divide the sandwich into three pieces. Say that it does not matter if the pieces of sandwich differ in shape and/or size. The first cut should produce two pieces, so avoid cutting a 'wedge' in the sandwich.

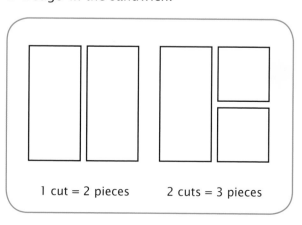

1 cut = 2 pieces   2 cuts = 3 pieces

Drawing together

● Pick one group to demonstrate how they acted out the problem. Check that all of the groups agree that you need to make two cuts to divide the sandwich into three pieces. Ask the children to think about what would happen if you wanted to divide the sandwich between a different number of bears. Would you have to do more or less cuts to divide the sandwich if there were two bears or four bears? How do you know?

● Explain that you would like each group to work together to investigate how many cuts they need to make to divide a sandwich between different numbers of bears. Provide each group with several squares of paper so that they can solve the problem practically by cutting the 'sandwich' into the required number of pieces. Give each group a copy of photocopiable page 53 to record their results.

● Remind the children that this is a real exercise and that the rule is that each cut has to result in a new piece and each piece has to be cut separately. For example, once a piece is cut in two, you will need to make one cut in each piece to make four pieces; you can only pick one piece up at a time to cut it.

Support

Work through the first few sections of the table together. Demonstrate to the children how to use the paper and scissors to solve the problem practically.

Extension

Challenge more able children to devise their own results table.

Further idea

Ask the children to solve the same problem in a different context, such as cutting up a cake for guests at a party.

Royal tea party

Setting the context

Cinderella and Prince Charming are having a party to celebrate their first wedding anniversary. They have invited their closest friends to come to the party. The preparations are underway, but in the dining room, the royal servants have a problem. At the last minute Cinderella and Prince Charming have decided that they want all the guests to sit together around one long table. The servants cannot work out how many small tables they need to push together to make a table which is big enough for all of the guests to sit around.

Problem

Cinderella and Prince Charming have 22 guests coming to their party. How many tables will the servants need to push together in a row so that everyone at the party will be able to sit around the table for tea?
● Four people can sit around one square table;
● Six people can sit around two square tables that have been placed together side by side.

Objectives

To solve mathematical problems and puzzles. To devise and use an appropriate problem-solving strategy.
To identify, describe and continue patterns in a sequence of numbers.

You will need

Several square carpet tiles or mats; several small plastic or cardboard squares; counters; small people figures; photocopiable page 54 (adjust the number of guests before copying the page, to suit the children's abilities.)

Preparation

Choose different children to play the parts of the characters in the story.

Solving the problem

● Introduce the problem by asking the children to act out the story ('Setting the context' above).
● Ask the servants to place a carpet tile/mat on the floor to represent a table. Choose four guests to sit around the 'table'. Repeat with a second table. Count how many guests are able to sit at two tables (eight).
● Tell the children to watch what happens

when the servants push the two tables together side by side. Ask: *How many children can sit down now?* (Six) *What has happened to the places where these two children were sitting?*
● Display the problem on the board. Ask the children to describe, in their own words, what the problem is asking them to find out. How might the children solve the problem? Discuss the children's ideas. What sort of equipment do the children think they could use to help them represent the tables and the guests at the party (such as small plastic squares and counters)?
● Demonstrate some of the strategies that the children suggest, such as drawing the tables on squared paper, joining small squares together in a row.

● Explain that you would like the children to work in small groups to investigate the problem. Describe how you would like the children to be systematic in their approach to solving the problem and recording what they find out. Tell them to start by counting how many guests can sit around one table, then two tables and so on. Give each group a copy of photocopiable page 54 to record their results.

● Give the children sufficient time to solve the problem. Extend the activity for groups who finish quickly (see below). Ask questions to guide less confident children into making appropriate decisions about the strategies they are going to use.

● As the children are working, encourage them to look for patterns and relationships between the number of tables and the number of people who can sit down. Ask: *How many places are created each time you add a table? What do you notice about the numbers in this column? Can you predict what the next number will be?*

Drawing together

● Invite the children to say how many tables the servants need to push together to seat all the guests at the party. Let different groups demonstrate how they solved the problem.

● With the children's help, fill in an enlarged copy of the results table on photocopiable page 54. Pose simple questions that encourage the children to look for patterns and relationships between the numbers in the table. Ask: *What do you notice about the numbers in this column? Are the numbers getting smaller or bigger? What is the difference between the numbers?* Establish that the numbers in the right hand column are all even and have a difference of two between them.

● To check which children have understood the 'add 2' rule, ask everyone to predict how many guests could sit around a certain number of tables. Check the answer using the practical apparatus.

Support

Show children who are unable to devise their own problem-solving strategy a particular method they could use to explore the problem, such as using small plastic squares and counters to represent the tables and guests.

Extension

At the last minute Cinderella has decided that she does not want the tables to be set out in a row. Ask: *Can you find a different way of arranging the tables so that there are enough places for all of the guests who are coming to the party?*

Further idea

Challenge the children to investigate the pattern of numbers that would be created using triangular-shaped tables.

Stuck on the roof

Setting the context

It is a beautiful summer's day. The sun is shining and the birds are singing, but all is not well down on Farmer Green's farm. George the cat is stuck on the roof of the barn. The other animals have tried to persuade George to climb down but he is too frightened. Luckily Farmer Green has had an idea. He is going to fetch some bales of hay from the hayloft and use them to build a staircase so that he can climb up onto the roof of the barn and rescue George himself.

Problem

How many bales of hay does Farmer Green need to fetch from the hayloft so that he can make a staircase with seven steps to rescue George?

Objectives

To solve mathematical problems and puzzles. To look for patterns in number sequences. To explain methods and reasoning.

You will need

Small wooden building bricks; cubes; a toy farmyard set (if available); photocopiable page 55.

Preparation

Set up the farmyard scene to illustrate the problem (see 'Setting the context' above).

Solving the problem

● Tell the children about the drama that is taking place on Farmer Green's farm. Use the figures in the farmyard set to depict the scene. Alternatively, choose volunteers to act out the story. Ensure that the children know what hay bales are and what they look like.

● Read the problem to the children and ask them to explain, in their own words, what they need to find out. Ask the children to suggest what you could use to represent the hay bales (such as wooden blocks or cubes).

● Place one hay bale (wooden block) on the ground. Using the toy farmer, illustrate how this represents one step.

● Next, demonstrate how to add bales to construct a second step. Ask: *How many hay bales did I add to make the second step? How many bales are needed altogether to make a staircase with two steps?* Invite one of the children to come and add a third step to the staircase. Pose the questions again.

● Now explain that you would like everyone (individually or with a partner) to solve the problem of how many hay bales Farmer Green needs

to fetch from the hayloft by actually building a staircase with seven steps themselves. Instruct the children to start with one step and then make the staircase grow, one step at a time. Ask: *How can you keep track of the number of hay bales needed to add to each additional step?*

● Show the children the photocopiable page. Demonstrate how to complete the table by recording, in the middle column, the number of cubes they add to make each additional step; and recording, in the right column, the total number of hay bales that have been used to make each staircase.

● Allow the children sufficient time to work through the problem using small blocks or cubes to represent the hay bales. Ask questions to support and develop the children's understanding of the problem and questions that encourage them to look closely for patterns. Ask: *How many bales did you need to add to build the last step? How many bales will you need to build the next step? How do you know?*

● Show children who complete the task quickly how to record the staircase pattern on squared paper.

Drawing together

● Share the children's solutions to the problem. Stick an enlarged copy of the photocopiable sheet on the board. Fill it in with the children's help. Ask the children to look closely at the numbers in the middle column of the table and describe the number pattern (+1). Help them to deduce from this that to work out how many bales are needed to build a staircase with eight steps, they simply need to add eight to the total number of bales that were needed to construct the staircase with seven steps.

● Test this rule with the children. Ask the children to predict how many bales Farmer Green would need to make a staircase with eight steps. Check their answers by adding an eighth step to the staircase and counting up the total number of cubes used to build the staircase.

Support
Show the children how to build the staircase using a different colour for each new step added. This may make it easier for some children to identify and describe the pattern of the growing staircase.

Extension
Help the children to see that by adding together consecutive numbers they can work out how many bales they would need to build a staircase with any number of steps. For example, 1 + 2 + 3 + 4 = 10, which is the number of bales that Farmer Green needs to build a staircase with four steps. Ask the children to test this rule.

Further idea
Repeat the same problem in a different context. Investigate how many blocks the prince needs to make a staircase with six steps to rescue the princess from the tower.

Jasper's beanstalk

Setting the context

The story *Jasper's Beanstalk* by Nick Butterworth (Hodder Children's Books) provides an ideal context for this problem-solving lesson. Alternatively, make up your own story about a child who plants a seed and watches it grow.

On Monday, Jasper the cat finds a bean, and on Tuesday he plants it. He then spends the next week watching and waiting until, finally, a beanstalk begins to grow.

Problem

Jasper's bean grew 1cm on a Friday. After that, the beanstalk grew 2cm taller each day. When did the bean reach 13cm tall?

Objectives

To solve mathematical problems and puzzles.
To devise and use appropriate problem-solving strategies.
To continue a number sequence according to a given rule.

You will need

Jasper's Beanstalk by Nick Butterworth (Hodder Children's Books); a bean; a pot; a watering can and other props needed to role-play the story; cubes; counters; rulers; paper and pencils.

Solving the problem

● Begin the lesson by reading the story *Jasper's Beanstalk* to the class. Choose one of the children to be Jasper and to act out each section of the story using the props you have gathered.

● Read the problem to the class. Draw and label a picture of the beanstalk on Friday when it is 1cm tall, and a picture of the beanstalk when it is 13cm tall to help the children to visualise the problem. Ask the children to describe, in their own words, what the problem is asking them to find out.

● Help the children to identify the information they need to be able to continue the pattern by asking questions such as: *How tall was the beanstalk on Friday? How many centimetres taller does the beanstalk grow each day? How tall will the beanstalk be on Monday? How do you know?*

Write the following questions on the board: *How are you going to solve the problem? What equipment will you use? How will you record what you have found out?*

Ask the children to spend a few minutes answering the questions with a partner and then discuss the children's responses as a whole class. This provides the children with an opportunity to think and talk about how they are going to tackle the problem, as well as helping you to identify any children who have not fully understood what the problem is asking them to do.

BEANSTALK 1CM   BEANSTALK 13CM

● Give the children sufficient time to solve the problem using their chosen method. Ask them to record what they are doing on paper (for example, in a list or table or pictorially).

● Provide support and guidance to the children as they work. If necessary, demonstrate how the problem can be modelled using cubes. Place one green Multilink cube on the table to represent the beanstalk on Day 1. Record this on the board as follows:

Friday   1cm   ☐

● Remind the children that the beanstalk grows 2cm each day. Ask one of the children to add more Multilink cubes to show what the beanstalk looked like on Saturday. Again, model how to record this information clearly:

Saturday   3cm   ☐
                                ☐

● Ask the children to continue the 1 + 2 + 2 pattern using cubes until the beanstalk is 13cm tall (on Thursday).

Drawing together

● Choose different pairs of children to present their work to the rest of the class. Ask them to explain how they solved the problem. Model some of the strategies the children have used on the board. These might include using a ruler to draw an accurate picture of the height of the beanstalk each day; recording the information in a list, using cubes/counters to continue the pattern; and writing a number sentence such as the one below:

1   +2   +2   +2   +2   +2   +2
F   Sa   S   M   T   W   Th

● Invite the children to say which of these strategies they think is the most efficient and why.

● Work together to devise a rule for the number sequence generated in this pattern. Ask the children to calculate how tall the beanstalk will be after one more

day. Assess which children are able to use the rule to predict the next few numbers in the sequence correctly (ask them to show their answers on a number fan).

Extension
Ask: When will the bean reach 20cm? This extension activity challenges more able children to recognise and explain that by starting from 1cm and adding 2cm each time they will only ever land on odd numbers and therefore it is not possible to say when the bean will reach 20cm.

Further idea
Set similar challenges in different contexts, such as asking: *It costs £5 to get into the funfair. Each ride costs £1. Henry has £13 to spend. How many rides can he go on?*

Fruity pyramid

Setting the context

This activity uses the story *Handa's Surprise* by Eileen Browne (Walker Books) as a starting point. Handa is on her way to take a basket of delicious fruits to her friend in another village. Her walk takes her past a variety of animals who think the fruit looks very exciting and they help themselves. When she arrives at the next village she has a big surprise. The basket is full of tangerines!

Problem

Handa stacks the tangerines in her basket into a pyramid. The bottom layer of the pyramid looks like this if you are looking down on it (see photocopiable page 56).

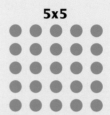

5x5

The pyramid has four more layers.
● How many tangerines are in each layer?
● How many tangerines are in Handa's pyramid altogether?

Objectives

To solve mathematical problems and puzzles.
To explore number patterns.
To name and describe 2D and 3D shapes using appropriate mathematical vocabulary.

You will need

Handa's Surprise by Eileen Browne (Walker Books); a basket of small oranges/balls to represent Handa's basket of tangerines; a pyramid shape; some dough (preferably orange); photocopiable page 56; a pot of counters; Blu-Tack; five pieces of paper (15cm x 15cm) for each group.

Preparation

Enlarge a copy of photocopiable page 56 and stick it on the board.

Solving the problem

● Introduce the activity by reading the story *Handa's Surprise* to the children. Show the children the basket of 'tangerines'. Ask them to suggest different ways that Handa could stack the tangerines.
● Ask the children to sit in a circle. Using the enlarged photocopiable from page 56 on

the board, read the problem and talk it through with the children.

● Remind the children of what a pyramid looks like by passing a pyramid shape around the circle. Ask the children to describe the properties of the pyramid using appropriate mathematical vocabulary.

● Pick a child to construct the first layer of the pyramid using the tangerines in the basket. Ask the children to say what shape it is and to count how many oranges were used to make it.

● Remind the children that Handa's pyramid has four more layers. Show the children the pyramid shape again. Stress that the pyramid is widest at the base and gradually narrows to a point at the top.

● Ask the children to predict whether the next layer of the pyramid will have more or less tangerines than the first.

● Tell the children that you would like them to help you build the next layer of the pyramid. Demonstrate how one tangerine rests on top of four others. Invite different children to come and add a tangerine until the second layer is complete. Count how many tangerines were used to build the second layer.

● Show the children how to record what this layer of the pyramid looks like by sticking counters onto the whiteboard to represent the arrangement of the tangerines.

● Ask the children to describe any similarities or differences they notice between the first two layers of the pyramid. How many more oranges were there on the first layer compared to the second layer? (Nine). Ask: *What do you think the next layer will look like? What shape do you think the next layer will be?*

● Organise the children to work in small groups. Give each group a large ball of dough, a copy of photocopiable page 56, five pieces of paper, a pot of counters and some Blu-Tack. Instruct them to build the complete tangerine pyramid, one layer at a time, using small balls of dough to represent the tangerines and the Blu-Tack to keep them in place. Ask the children to record what each layer of the pyramid looks like on a separate piece of paper, using counters.

● As the children are solving the problem, circulate around the groups. Encourage the children to look for and describe patterns in their results. Ask questions such as: *What are you trying to find out? What do you notice about the number of tangerines you need to build each layer? How many tangerines do you think you will need to make the next layer?*

Drawing together

● Look at the pyramid shapes created by the children. Refer back to the original problem. Ask each group to say how many tangerines they think Handa used to build her pyramid.

● Identify a group who have recorded the arrangement of tangerines in each layer accurately. Ask them to stick their five pieces of paper on the board. Count how many tangerines are in each layer of the pyramid. Can the children predict how many tangerines Handa would need to build another layer on the bottom of her pyramid?

Support
Build a complete pyramid with the children before asking them to construct their own tangerine pyramid using balls of dough.

Extension
Ask: *If Handa added another layer onto the bottom of her tangerine pyramid what would it look like? How many tangerines would she need? How many tangerines would she need for eight layers?*

Further idea
Set a similar problem in a supermarket context, for example, cans of cat food stacked in a big triangle or pyramid shape.

Take the train

Teddy bears' picnic

- How many cuts would you have to make so that each bear can have a piece of sandwich?

- It does not matter if the pieces of sandwich are different shapes or sizes.

- Find out how many cuts you would need to make to share the sandwich between different numbers of bears.

Number of bears	Number of cuts

Royal tea party

- Four people can sit around one square table.

- Six people can sit around two square tables that have been placed together side by side.

- Cinderella and Prince Charming have _____ guests coming to their party. They want everyone to sit together on one long table. How many tables will the servants need to push together in a row so that everyone at the party will be able to sit around the table for tea?

Number of tables	Number of people who can sit at the table
1	
2	
3	

■SCHOLASTIC
www.scholastic.co.uk

Stuck on the roof

- Oh no! George the cat is stuck on the roof of the barn and he cannot get down!

- How many bales of hay does Farmer Green need to fetch from the hayloft so that he can make a staircase with 7 steps and rescue George?

- He needs 1 hay bale to make 1 step.

- He needs 3 hay bales to make 2 steps.

Number of steps	Number of bales added	Total number of bales
1	1	1
2	2	
3		
4		
5		
6		
7		

Tangerine pyramid

- Handa stacks the tangerines in her basket into a pyramid.

- The bottom layer looks like this.

- The pyramid has four more layers.

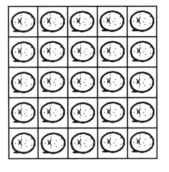

- How many tangerines are in each layer?

- How many tangerines are in Handa's pyramid altogether?

- Draw each of the other layers of the pyramid.

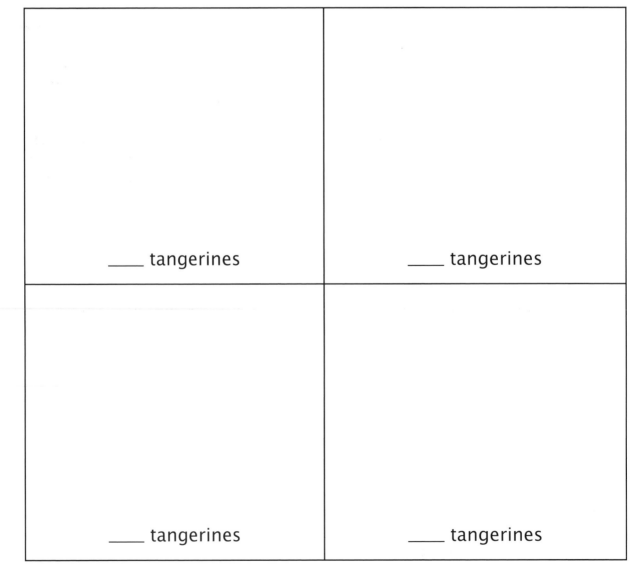

_____ tangerines

_____ tangerines

_____ tangerines

_____ tangerines

■SCHOLASTIC
www.scholastic.co.uk

Chapter Three

Diagram problems and visual puzzles

The problems in this chapter can be classified as either diagram problems or visual puzzles. The visual, practical nature of both these types of problem make them particularly engaging for young children whatever their preferred learning style is. All of the problems are introduced using either a picture or diagram or practical materials as the starting point. Children are encouraged to think creatively and flexibly when exploring different strategies that can be used to solve problems.

As with any type of problem, children need to be taught a range of appropriate strategies for solving 'diagram problems and visual puzzles' and they need to learn how to be systematic in their approach when solving them.

The activities in this chapter are designed to teach children to:
- read the problem carefully and identify the 'rules' of the problem;
- explore different strategies and make decisions about how to record outcomes;
- select and use appropriate equipment to solve the problem;
- use different recording strategies to keep track of what they are doing, for example, drawing pictures (see 'Squashed sheep', pages 68-69) or using a tally (see 'Help me!', pages 70-71);
- try different approaches and persevere until they have successfully completed a task (see 'Mrs Muddle' pages 60-61 and 'Patchwork quilt', pages 64-65);
- use a systematic approach, for example, when counting shapes (see 'Kandinsky', pages 72-73).

Rakhi bracelets

Setting the context
The Hindu festival of Raksha Bandhan takes place in August. It is a festival which celebrates the relationship between brothers and sisters. Traditionally, during the festival, girls tie a rakhi (a bracelet made from interwoven red and gold threads) around their brother's right wrist and pray for his well-being. Brothers usually give a gift in return and pledge to take care of their sisters. Nowadays, rakhis are also often adorned with different coloured stones and beads.

Problem
Design and make a rakhi bracelet with three green, three blue and three red beads. Beads of the same colour must not be placed next to each other.

Objectives
● To solve mathematical problems or puzzles.
● To explain methods and reasoning.

You will need
A copy of photocopiable page 74 for each child; counters and small coloured circles or sequins (blue, green and red); glue; different-coloured wool; thin card; circle templates; nine large paper circles (three red, three green, three blue); a hoop.

Preparation
Prepare a rakhi bracelet following the instructions on photocopiable page 74.

Solving the problem
● Tell the class about the festival of Raksha Bandhan. Show the children a picture of a rakhi bracelet. Ask: *What materials do you think the bracelet is made out of? Can you describe the pattern on the bracelet?*
● Show the children the rakhi bracelet you have made. Choose a boy and a girl to come out to the front. Help the girl to tie the bracelet around the boy's right wrist to demonstrate the Hindu custom that takes place during the festival of Raksha Bandhan.
● Tell the children that you would like them each to design and make a rakhi bracelet. Explain that there are some special rules to follow when they are designing the pattern on their bracelet.
● Read the problem and ask the children to explain, in their own words, what they have got to do and what the special rules they must follow are (they must use nine beads,

three each of red, green and blue; beads of the same colour must not be next to each other).

● Draw a large circle or stick a hoop on the board. Pick a child to arrange three large paper circles around the inside edge of the circle (see below). Refer back to the original problem. Ask the children to say whether they think this large design follows the rules of the problem. Why? If necessary, continue to rearrange the circles until an appropriate solution is found. Explain that there are several solutions to the problem.

● Now, outline the task. First, you would like the children to design a bracelet by placing coloured counters onto the bracelet shape on the photocopiable page; it must be different from the one on the board. They should then check that the bracelet they have designed meets the rules of the task. Finally, they should follow the instructions on the photocopiable sheet to make the bracelet they have designed.

● If you wish to make the activity more challenging, and encourage dialogue between the children, organise them to work in groups of three. Stipulate that each child in the group must design a different bracelet.

Drawing together

● Ask the children to sit in a circle. Spread the bracelets out on the floor. Look closely at several of the bracelets. Ask the children to remember what the rules of the problem were. Did the children who designed these bracelets follow the rules?

● Encourage the children to look closely at all of the bracelets. Work together to sort the bracelets into piles according to their design.

● Ask: *How many different bracelets have we designed? Do you think we have found all the designs you could make with this number of beads? Do you think we would have found more or less designs if we had used four*

beads of each colour? Why? What about if beads of the same colour were allowed to touch?

Support

Work alongside less able children while they are designing their bracelets. Help them to check whether their design meets the rules of the task. If the design is unsuitable, make sure the children understand why and encourage them to try again.

Extension

Challenge more able children to investigate how many different bracelet designs it is possible to make, following the rules of the task.

Further ideas

● Design a flower garland using different types and colours of flowers.

● Tell the children that no one is allowed to sit next to someone who has the same colour hair as them! Challenge the class to find different ways that they can line up following this rule.

Mrs Muddle

Setting the context

Mrs Muddle is a baker. She works in a busy bakery making bread and cakes. Unfortunately, as her name suggests, Mrs Muddle often muddles things up! Mrs Muddle would like the children to help her sort out her latest muddle. This morning a lady phoned the bakery to order five cakes for her daughter's fifth birthday party. She asked for each cake to be decorated with five cherries. Unfortunately Mrs Muddle forgot this and put a different number of cherries on each cake. Mrs Muddle has not got time to make any more cakes, but she has a plan. She has cut each cake into four pieces and is going to try and put the pieces back together in a different way so that there are five cherries on each cake.

Problem

Arrange the pieces of cake to make five cakes, each with five cherries on top.

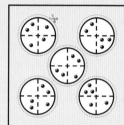

Objectives

To solve mathematical problems or puzzles. To understand the operation of addition and the related vocabulary.

You will need

Five paper plates; red tissue paper; an apron and/or chef's hat; cut-up sets of the 20 cake pieces from photocopiable page 75 for each group.

Preparation

Paint or colour the paper plates brown to represent cakes. Stick balls of tissue (cherries) onto the cakes (see diagram). Cut each cake into quarters.

Solving the problem

● Put on the apron and hat. Greet the class in role as Mrs (or Mr) Muddle. Tell them about yourself. Explain that you would like the children to help you sort out your latest muddle, otherwise you might lose your job at the bakery!

● Invite the class to sit in a circle. Spread the 20 pieces of cake on the floor. Ask different children to make a cake with five cherries on top. Make sure that all the children understand that there are lots of ways of making five.

● Explain to the children that you would like them to try and use all the pieces of cake to make five cakes, each with five cherries on the top. Pick different children to make a cake until the task has been completed successfully, or you are left in a position where the remaining pieces of cake cannot be put together to make a cake with five cherries. If this happens, praise the children for their efforts but explain that they must break up some or all of the cakes and experiment with different ways of putting the pieces of cake together.

● Once the five cakes have been correctly assembled (there is more than one solution) thank the children for their help.

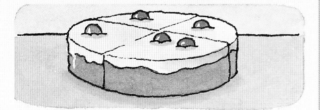

● Read the problem with the class and explain how the children have solved this problem using the plates and tissue paper cherries. Explain that the children are now going to get the opportunity to find another way of solving the same problem, this time using the pieces cut from photocopiable page 75. Organise the children to work in pairs or small groups. Give each group a set of the 20 pieces from photocopiable page 75. Ask the children to reassemble the pieces of cake to make five cakes, each with five cherries on top. Explain that there are a number of possible solutions to the problem.

● Offer the children support and encouragement as they are working. Encourage them to try different approaches and persevere until they have successfully completed the task. If a group gets to the last cake and has the wrong number of cherries left, demonstrate to the whole class how it may be easier to try and swap one or two bits of the puzzle round rather than breaking up everything they have done so far.

● With about ten minutes of the lesson to go, tell each group to stick their puzzle pieces down on a sheet of A4 paper.

Drawing together

Hold up one group's solution to the problem. Ask: *How can we check that this group have solved the problem correctly? Has anyone found a different solution to the problem?*

Support
Ask less-able children to rearrange the puzzle pieces to make as many cakes as they can with five cherries on top.

Extension
● Challenge the children to investigate how many other solutions they can find.
● Prepare a second puzzle for children who finish quickly. Adapt the photocopiable sheet by drawing an extra cherry on each cake. Challenge the children to make five cakes with six cherries on the top.

Further ideas
● Set the same problem in a different context. For example, put six fish in each pond or six pieces of topping on each pizza.
● Give each group of children five paper plates and ask them to devise a similar problem for another group to solve.

Keep off!

Setting the context

Jake and Adele (insert the names of two children in your class) were both feeling grumpy. It was a very hot day and they were stuck inside the classroom practising some really hard spellings. Suddenly Adele shouted, 'Oi, move your arm! It's on my side of the table!'

'No it's not! That's my side,' shouted Jake angrily. 'And anyway your pencil case is on my side of the table!' he moaned, shoving it onto the floor.

'Now, now stop all this noise,' said Mr Patrick (insert your name) their teacher. 'I've got an idea. I will draw a line to divide your table in half. Then you will each have exactly the same amount of space and there will be no more arguing!'

Problem

Can you divide the table so that the two children each have exactly the same amount of space? Find different ways to divide the table into two equal parts.

Objectives

To solve mathematical problems or puzzles.
To divide shapes into two equal parts through folding and cutting.
To explain methods and reasoning.

You will need

A table and two chairs; a pencil case; scissors; small rectangular pieces of paper; glue.

Preparation

Put a table and two chairs at the front of the room.

Solving the problem

● Choose two children to sit at the table. Explain that you are going to tell a short story and that you would like them to act out the parts of the children in the story. Discuss the problem that is introduced by the story. Ask the children to explain, in their own words, what the teacher has decided to do to try and stop the two children squabbling.

● Display the problem and read it to the class. Discuss the meaning of key vocabulary: divide, equal, half. Question the children to check that they understand exactly what the problem is asking them to do.

● Tell the children that you would like them to pretend that they are the teacher in the story. Stick a rectangular piece of paper on the board. Ask the children to imagine that the paper represents the top of the table at which the two children are sitting. Invite one child to come and draw a line on, or fold, the paper to divide it into two equal parts. Ask the rest of the class to indicate whether or not they think that the table has been split fairly. Can the children suggest possible strategies to check whether or not the two parts are equal?

● Demonstrate how to check whether the rectangle has been divided into equal parts. First cut along the pencil line. Then place the two pieces of paper on top of each other and hold them up to show the group. Ask: *Are the parts equal? How do you know?*

● Organise the children to solve the problem in pairs. Provide rectangular pieces of paper, pencils and scissors, for the children to test their ideas.

● Instruct the children to keep a record of the different solutions to the problem. Either let the children devise their own method of recording or, alternatively, specify exactly how you want them to record their solutions, for example, drawing pictures or gluing the divided rectangles onto a piece of paper.

● Encourage the children to discuss their ideas and methods of recording as they tackle the problem. Ask them to show you any solutions they have found. (You may need to demonstrate that in some cases when you put the two parts of the rectangle together, to check that they are equal in size, it is necessary to rotate or flip one of the pieces over.)

Drawing together

● Share the different solutions that the children have found to the problem. Count how many different ways the children have found to divide the table into two equal parts.

● Explain that providing you fold through the centre point of the rectangle it will always divide into two equal parts.

● Finish the lesson by concluding the story you used to introduce the problem. Allow one child to be the teacher and draw a chalk line across the table to divide it into two equal parts. Now the two children can keep to their own part of the table and stop squabbling!

Support

Some children may need help to fold and cut the rectangles accurately.

Extension

Ask the children to investigate different ways of dividing a table into four equal parts.

Further ideas

● Repeat the same problem in a different context. For example, ask children to find ways of splitting a sandwich into two equal pieces.

● Ask the children to devise their own problem based on two friends who are squabbling about getting their fair share of something (such as the space in a sand pit). Choose one of the problems to solve collectively.

Patchwork quilt

Setting the context
Read *Granny's Quilt* by Penny Ives (Puffin). Discuss what a patchwork quilt is and why quilts are treasured by their owners, especially if they are hand-made.
Tell the children that Granny is going to make a small patchwork quilt for her granddaughter. Her granddaughter is 12 years old and so Granny is going to try and create a special design for the quilt based on the number 12.

Problem
Can you arrange the nine shape patches so that the number of sides of the shapes in every row and column on the patchwork quilt add up to 12?

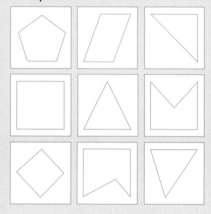

Objectives
To solve mathematical problems or puzzles. To count the number of sides on a variety of 2D shapes.

You will need
A patchwork quilt or a picture of a patchwork quilt; a copy of photocopiable page 76 for every child in the class; scissors; glue.

Preparation
Enlarge, colour and laminate a copy of the shape patches from photocopiable page 76.

Solving the problem
● Set the context for the lesson using the introduction suggested above. Show the children examples of patchwork and explain how patchwork is made. Give the children an opportunity to describe other examples of patchwork they may have seen.
● Show the children the laminated shape cards. Explain that these represent the pieces of material that Granny is going to use to make the patchwork quilt.
● Invite different children to name the shapes on the cards. As a class, count the number of sides each shape has. Remind the children of a strategy they can use to ensure that they count the sides correctly – putting a finger on the first side they count and then counting the sides of the shape in order.
● Choose a child to arrange the shape cards into a 3x3 square. Explain to the children that Granny is planning to make the quilt by sewing the nine patches together in a special arrangement. She has had an idea for a design based on the number 12, but needs the children's help.
● Read the problem. Make sure that the children understand what the problem requires them to do and that they understand the meaning of the terms 'row' and 'column'.
● Give each child a copy of the photocopiable page. Instruct them to cut out the shape cards and then try to arrange them

on the grid so that the number of sides of the shapes in each row and column adds up to 12.

● Work alongside less able children. If necessary, teach the children a strategy for adding up the total number of sides of the shapes in a single row or column. For example, count and record the number of sides of each shape and then add the three numbers together. For example, $4 + 4 + 4 = 12$.

● Offer the children support and encouragement as they are working. Encourage them to experiment with different arrangements and persevere until they have successfully completed the task.

● If the children finish the activity quickly, challenge them to find different ways of solving the problem.

Drawing together

● Ask the children to stick down the shape cards on the 3x3 grid to record a solution to the problem.

● Stick one of the children's completed quilt designs on the board. Ask them to explain the strategy they used to solve the problem to the rest of the class.

● Invite the children to help you check whether the design meets all the rules of the problem. Refer back to the original problem to remind children what these rules were. Ask children to add up the number of sides of the shapes in each row and column in turn, to check they add up to 12.

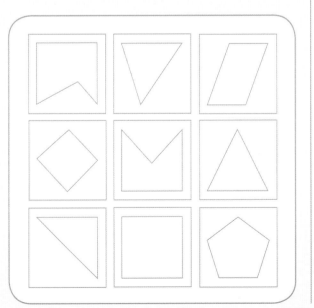

● Invite the children who have found a different solution to the problem to come out and show their work. Count how many different solutions the children have found altogether.

● Tell the children that you are going to send the completed designs to Granny so that she can look at them all and decide which one she is going to make.

Support
Simplify the activity so that the children only have to make the number of sides in each *row* add up to 12.

Extension
Challenge the children to arrange the shape patches so that the number of sides in every row, column and *diagonal* add up to 12.

Further idea
Give the children a blank 3x3 grid and a variety of small shape templates. Instruct them to design a patchwork quilt of their own. State the total number of sides that there should be in every row and column. Tell the children to cut up their patchwork quilt and give it to someone else in the class to put back together.

Adventure playground

Setting the context

Little Bear was really excited. It was his birthday and his dad had promised to take him and some of his friends to play in the park. Little Bear could hardly wait! Little Bear loved climbing and was really looking forward to having a go on the climbing equipment in the new adventure playground. When they arrived at the playground Little Bear and his friends were really disappointed. All the climbing frames looked exactly the same. Can you help redesign the adventure playground to make it a more exciting place for Little Bear and his friends to play?

Problem

How many different pieces of climbing equipment can you construct by fitting five Multilink cubes together face-to-face.

Objectives

To solve mathematical problems or puzzles. To investigate different shapes that can be made by fitting five cubes together face-to-face.

You will need

A stick of five Multilink cubes for every child (the sticks of cubes can be different colours, but the cubes in each stick should all be the same colour); sheets of 1cm squared paper; plain A4 paper; crayons; Compare Bears.

Preparation

Organise the children so that they are sitting in a circle. Spread the sticks of Multilink out on the floor and position the Compare Bears on and around the cubes.

Solving the problem

● Introduce the problem by telling the story of Little Bear and his friends.

● Direct everyone in the circle to take a stick of Multilink, change it into a different shape and then return it to the centre of the circle. Explain that the cubes must be joined together, side by side, so that the completed shape is the thickness of one cube. (Some children may need to use a small piece of Blu-Tack to stop the shape they have made toppling over.)

● Praise children who follow your instructions correctly. Say that the adventure playground looks more interesting already.

● Ask the children to say what all of the shapes have in common. Discuss and compare the range of shapes that the children have been able to construct simply by arranging the same number of cubes in different ways.

● Draw one of the cube shapes on the board. Ask the children to identify that shape on the carpet. Repeat for other examples.

● Draw attention to examples of shapes that look different, but are actually the same. Demonstrate how rotating or turning shapes may reveal two shapes as the same, although at first glance they may have appeared to be different.

● Tell the children that you would like them to design an adventure playground with lots of different pieces of climbing equipment. Emphasise that each piece of equipment must be made by fitting five cubes together

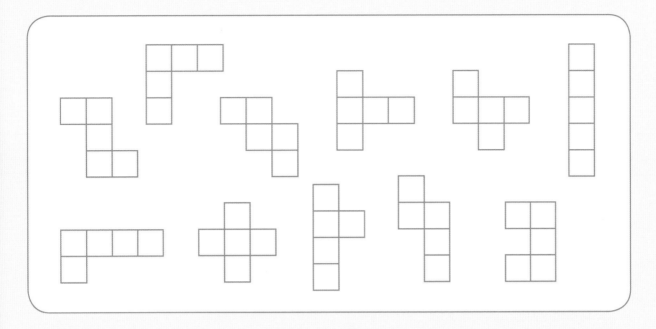

side by side. Ask the children to lay the piece they have made flat on the floor so that each cube touches the ground (this makes it easier to draw and record 2-D representations).

● Ask the children to suggest ways in which they could record all the different shapes they make. For example, freehand drawings or colouring shapes on squared paper.

● Organise the children to work individually or in pairs or small groups. Give each group or individual five Multilink cubes and a crayon. For recording purposes, provide half the groups with squared paper and the other half with plain paper.

● Observe and interact with the children as they are engaged in the practical task. Encourage them to check for repeats before recording each new shape. If necessary, remind some children to rotate and turn shapes when checking for repeats.

Drawing together

● Choose one group to explain, in their own words, what the problem was about and how they solved it. Ask each group to say how many different pieces of climbing equipment they made.

● Ask: *Why do you think I asked you to record your solutions? If I asked you to do this activity again would you choose squared paper or plain paper to record your shapes on? Why?* Discuss the children's responses.

● Organise the class so that they are sitting in a circle. Let each group, in turn, make a

different shaped piece of climbing apparatus and place it in the centre of the circle. Continue until the children have made all the different shapes it is possible to make with five cubes (there are 12, see above). Count how many different pieces of climbing apparatus there are altogether.

● Put the bears on and around the cubes. Congratulate the children on designing a much more exciting playground for the bears!

Support

Allow children who have difficulty recording their work to simply draw around each of the different shapes they make.

Extension

Pick two of the shapes you have made. How many different shapes can you make by joining these two shapes together?

Further idea

Set a similar problem, for example, ask: *What different shaped patios can you make by placing six square slabs together side by side?*

Squashed sheep

Setting the context
Farmer Green is very excited. Yesterday, Sally his favourite sheep, gave birth to four lambs. However, this has given Farmer Green a problem. The sheep pen on his farm is really too small for five sheep to live in. Sally and her lambs are very squashed and Farmer Green does not have enough money to build them a new pen. Farmer Green has had an idea. He hopes that by rearranging the existing fence panels to make a different shaped pen he may be able to create a larger space for his sheep to live in.

Problem
How many different shaped pens can you make using ten fence panels? Which pen has the most space for the sheep to move around in?

Objectives
To solve mathematical problems or puzzles. To explain methods and reasoning.

You will need
Lollipop sticks; ten small garden canes (all the same length); sheets of squared paper.

Preparation
Enlarge a sheet of squared paper and stick it onto the board.

Solving the problem
● Ask the class to sit in a semi-circle. Tell the children about Farmer Green's problem. Arrange the canes on the floor in the shape of the sheep pen that Sally and her lambs live in (see below). Count how many fence panels (canes) have been used to construct the sheep pen.

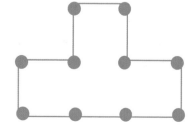

● Choose five children to be Sally Sheep and her lambs. Instruct them to stand inside the pen. Establish that it must be a real squash for all five sheep to live in the pen together.
● Draw the sheep pen on the enlarged sheet of squared paper. Demonstrate how the side of one square (on the paper) to represents each fence panel. Ask the children to count how many squares are enclosed by the fence. Explain that this space is called the area.
● Read and display the problem on the board. Ask different children to describe, in their own words, what they think the problem is asking them to do.
● Ask: *What equipment could you use to help you solve the problem? How could you record each of the different pens you make? How will you know which pen has the biggest area?*
● Ask a child to rearrange the sticks on the floor to make a different-shaped pen. Make sure that the children understand that it is important that the sticks join up to make an enclosed space so that the sheep cannot escape.

● Invite the sheep to move into the new pen. Do the sheep think they have more/less/the same amount of room to move around in? How can we check?

● Choose a child to draw the pen on the squared paper and then count how many squares have been enclosed by the fence. Ask: *Is this pen larger or smaller than the original pen? How do you know?*

● Organise the children to work in pairs or small groups. Give each group ten lollipop sticks to represent the fence panels and a piece of squared paper on which to record their work.

● Instruct them to investigate the different shaped pens they can make using ten fence panels.

● Observe how the children approach the task. Encourage them to discuss their work and interpret the information they have gathered in relation to the problem. Ask: *Which of the pens you have drawn do you think Farmer Green should make for his sheep? Why?*

Drawing together

● Refer back to the original problem. Remind the children that there were two parts to the problem. Ask them to count how many different shaped pens they have made and then decide which of the pens would give the squashed sheep the most space to move around in. Provide a sheet of plain paper for each group of children to draw their solution on.

● Give each group the opportunity to show their answer and explain why they think their pen is the best. Stick all the children's answers on the board. As a class, agree on the best one. Ask the children to explain their reasoning and how they know which is the biggest pen.

● Construct the chosen pen design on the floor with the canes. Tell the sheep to get into the pen. Take a photo of the sheep in their pen with a digital camera. Tell the children that you are going to send the picture to Farmer Green!

Support
Some children may find it difficult to record the pen shapes on 1cm-squared paper. Enlarge the squared paper so that the side of each square is approximately the same length as a lollipop stick, to make recording easier.

Extension
Ask the children to design a pen which encloses ten squares. Challenge them to find the smallest number of fence panels it is possible to do this with.

Further idea
Repeat the same problem in a different context. For example, make the biggest cage you can for an elephant using 18 fence panels; or, design the biggest flower border you can, using 14 fence panels.

Help me!

Setting the context

This lesson begins when the children find a mysterious box in the classroom. There is a letter attached to the box explaining who is inside the box and why they need the children's help.

Problem

Can you swap the witches and their cats to opposite sides of the grid (on photocopiable page 78)? A witch or cat can move into an empty space that is next to them or jump over one other witch or cat into an empty space.

Objectives

To solve mathematical problems or puzzles. To explain methods and reasoning.

You will need

A soft toy; a box; string; a copy of photocopiable page 78 for every child; one copy of the letter on photocopiable page 77; an envelope; three simple cardboard wands; seven hoops; an old pair of black tights.

Preparation

Put the toy in the box and use the string to tie the lid on the box. Put one copy each of photocopiable pages 77 and 78 into an envelope. Write 'PLEASE HELP ME!' on the front of the envelope and stick it on top of the box. Cut up the tights to make three tails.

Solving the problem

● Place the box in an obvious place in the classroom to ensure that it is quickly discovered by one of the children. Ask the children to speculate about what might be inside the package, who the letter might be from and how the box has come to be in their classroom.

● Open the letter and read it out. Discuss what the children need to do to break the spell the witch has put on the box and free whoever is trapped inside.

● Place seven hoops in a line on the floor. Pick three children to be the witches. Direct them to stand in hoops one to three. Choose three children to be the cats. Instruct them to

stand in hoops five to seven. To help the rest of the class remember who is a witch and who is a cat, give each of the witches a wand to hold and each of the cats a tail.

● Explain that the witches and cats are going to act out the problem. Before you begin, remind the group that the problem prescribes certain rules about how the witches and cats are allowed to move. Ask the children to say what these rules are. Write a list of the rules on the board so that the children can refer back to them at different points during the lesson.

● Act out the problem. Let the children take it in turns to suggest a move that one of the witches or cats could make. Encourage the rest of the group to refer to the board to

check that the moves are permitted within the rules of the problem. Continue until the witches and cats have been successfully swapped to opposite sides of the grid.

● Explain that there are a number of different solutions to the problem. Organise the children to work in pairs. Give each pair a copy of photocopiable page 78 and ask them to find different ways of solving the problem.

● Encourage more able children to try and devise a way of recording their moves. If you observe children making moves that are not permitted (such as jumping over two cats into an empty space) intervene and draw their attention back to the rules on the board.

Drawing together

● Ask one group to move the cats and witches in the hoops to demonstrate how they solved the problem. Instruct the rest of the class to watch carefully to check they adhere to the rules. Count how many moves the group use to solve the problem by keeping a tally on the board.

● Ask different pairs of children to explain how they recorded their moves. Discuss and compare the different methods the children have devised. Did they find it difficult or easy to record the moves?

● Finish the lesson by solving the problem on the mystery box. Involve the whole class in making decisions about which counter to move next. Choose a child to open the box and release the toy.

Page **71**

Support

Let less-able children focus exclusively on solving the problem. Do not ask these children to try and record their moves.

Extension

Challenge the children to find the least number of moves you can complete the task in.

Further idea

Set a similar problem, such as asking the children to swap two teams of five a-side footballers to opposite sides of the pitch. What is the least number of moves the children can do it in?

Kandinsky

Setting the context
The suggested starting point for this problem-solving lesson is the artwork of Wassily Kandinsky (1866-1944). Many of Kandinsky's paintings are abstract pictures made up of a variety of different shapes, colours and lines.

Problem
Look carefully at the overlapping shapes. How many triangles can you count?

Objectives
To solve mathematical problems or puzzles.
To use mathematical vocabulary to name and describe a variety of 2D shapes.
To work systematically and explain methods and reasoning.

You will need
A copy of photocopiable page 79 for each child; triangle templates; paper; coloured crayons; one or two Kandinsky prints such as 'Concentric Circles'; photocopiable page 80 (optional).

Solving the problem
● Display the Kandinsky pictures on the board. Allow the children to spend a few

minutes looking at the pictures.
● Ask: *What do the pictures have in common? Can you describe the shapes, lines and patterns that Kandinsky has used to create this picture?* Encourage the children to use appropriate mathematical vocabulary when naming shapes and describing their properties and position.
● Point to an example of where one shape has been painted so that it overlaps the edge of another shape. Can the children name the new shape that has been created between the overlapping shapes? Ask the children to study the pictures closely for other examples where two shapes overlap and create a new shape.
● Tell the children that you are going to draw a pattern, in the style of Kandinsky, on

the whiteboard. Copy the arrangement of triangles in the diagram (below). Ask the children to describe the pattern.

● Introduce the problem. Give the children, in pairs or individually, a few minutes to count the triangles. Then ask the children to show their answer on a number fan. Let children who counted the triangles correctly explain the strategy they used, for example, counting all the large triangles first.

● Explain to the children that you are going to teach them a strategy for keeping track of which triangles they have counted. Begin by counting the large triangles. Draw around the side of each triangle you count with a coloured pen and keep a tally. Then repeat exactly the same process for the small triangles. Ask: *Why do you think I coloured each triangle as I counted it?* If necessary, explain that it helped you to avoid missing any of the triangles out or counting any of the triangles twice.

● Give each child a copy of photocopiable page 79 and ask them to count how many triangles are in each of the patterns. Observe how the children approach the task and encourage them to talk about what they are doing and why. Praise the children who try to implement a systematic strategy for keeping track of the triangles they have counted.

● Support the children who find the task difficult. Help them to appreciate that it is important to work through the problem carefully and methodically. Remind them of the strategies they could use to do this. These children may find it easier to work in pairs, one of them marking the triangles which have been counted and their partner keeping a tally.

● Once the children have completed the task, suggest that they make their own

overlapping triangle pattern before tackling the other overlapping shape problems on photocopiable page 80.

Drawing together

● Pick different children to demonstrate and explain how they solved the problem. Ask: *How did you make sure you counted every triangle? Did you find it useful to keep a tally? Why?*

● Copy one child's overlapping triangle pattern onto the board. Invite the children to suggest appropriate strategies to use to make sure you count all of the triangles. Ask: *Which triangles shall we count first? What can we do to this triangle so that we know we have counted it?*

Support

Make a pattern by drawing around a selection of 2D shapes but do not overlap the shapes. Ask the children to count how many triangles they can see. Show the children how to keep track of which triangles they have counted by colouring them/covering them with a counter.

Extension

Draw a more complex pattern using a greater number of triangles. Vary the type, size and rotation of the triangles.

Further ideas

● Give children other opportunities to devise their own patterns of overlapping shapes using templates, stencils or a drawing program on the computer.

● Look at the artwork of Mondrian. How many four-sided shapes can you count?

● Ask the children to design their own problem for a partner to solve. They could use a single shape or different shapes.

Rakhi bracelets

- Use 3 red, 3 blue and 3 green counters to design a rakhi bracelet.

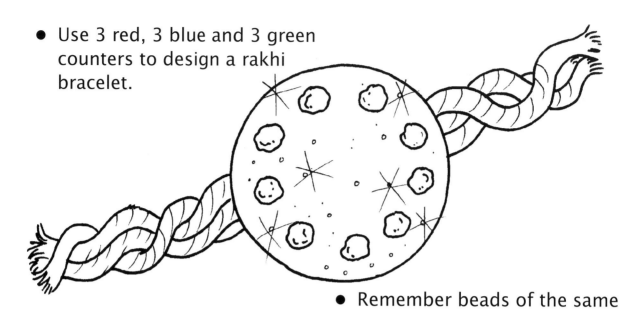

- Remember beads of the same colour must not be next to each other.

How to make your rakhi bracelet

You will need:
a small card circle (4-5cm in diameter)
glitter
several strands of wool
sticky tape
glue
9 small tissue paper balls (3 red, 3 green, 3 blue)

1. Twist the strands of wool together and stick them to the back of the card circle with a piece of sticky tape.

2. Decide the best pattern for the 3 colours of tissue paper balls for your rakhi bracelet. Draw and colour in your design on a piece of paper.

3. Stick the tissue paper balls onto the front of your card circle, following your design.

4. Finally, decorate your rakhi bracelet with glitter.

■SCHOLASTIC
www.scholastic.co.uk

Mrs Muddle

- Here is another batch of cakes from Muddle's Bakery. Oh dear, each cake is supposed to have five cherries on top but everything's in a muddle!

- Luckily the customer wants the cakes cut into quarters.

- Can you help out by rearranging the cakes so that each new cake has five cherries?

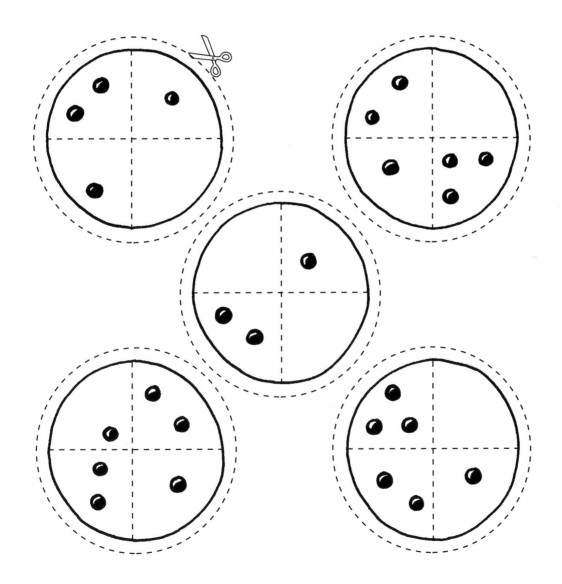

Patchwork quilt

- Can you help Grandma design a patchwork quilt for her granddaughter?

- Remember the number of sides of the shapes in each row and column must add up to 12.

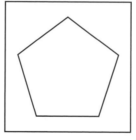 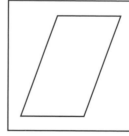 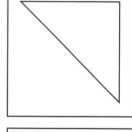

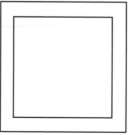 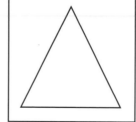 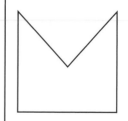

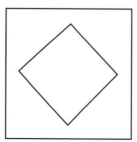 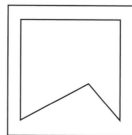 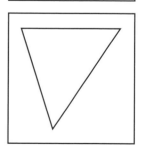

■SCHOLASTIC
www.scholastic.co.uk

Help me! (1)

To whoever should find this box.

Many years ago I was captured by a wicked witch. The witch put me inside this box and then put a spell on it to stop me from getting out.

Please break the spell and set me free. To break the spell, you must solve a problem:

Swap the witches and the cats to opposite sides of the grid.
You can move a witch or cat into an empty space that is next to them.
You can jump over one other witch or cat into an empty space.

Please help me!
From

Help me! (2)

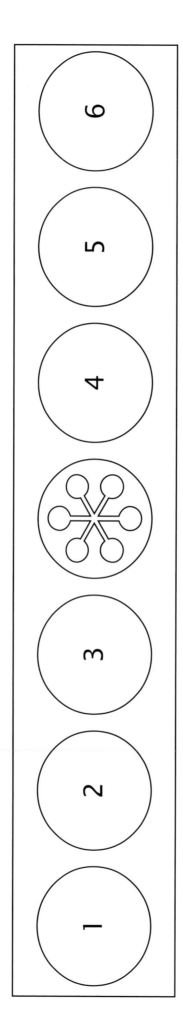

1	2	3		4	5	6

- Cut out the six counters.
- Place the witch counters in circles 1, 2 and 3.
- Place the cat counters in circles 4, 5 and 6.
- Swap the witches and the cats to opposite sides of the grid.

Remember the rules!

- You can move a witch or cat into an empty space that is next to them.
- You can jump over one other witch or cat into an empty space.

Kandinsky (1)

● How many triangles can you count in these pictures?

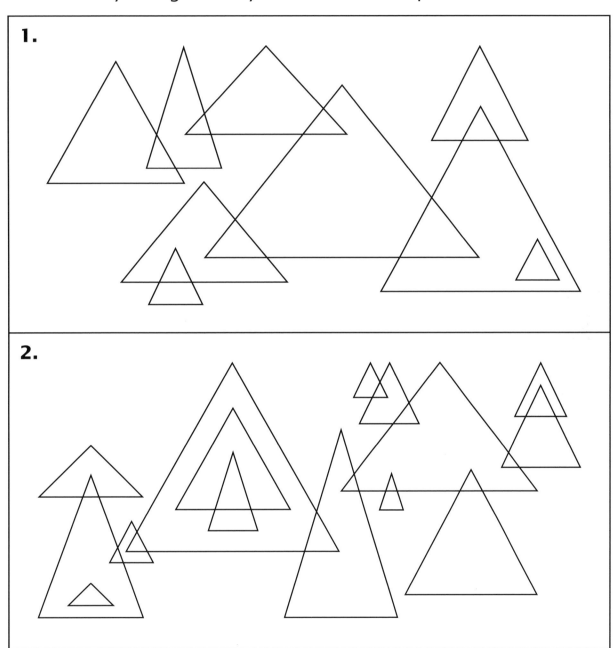

1.

2.

● Make a pattern of your own by drawing around a triangle template.

● How many triangles can you count in your picture? Ask your friend to count the triangles.

● Are your answers the same?

Kandinsky (2)

- How many squares can you count?

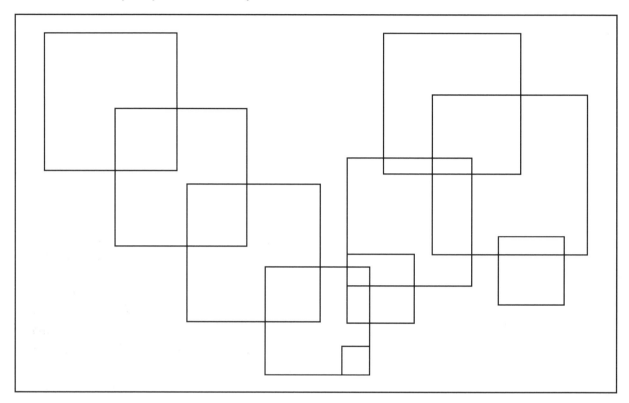

- How many squares, rectangles and triangles can you count?

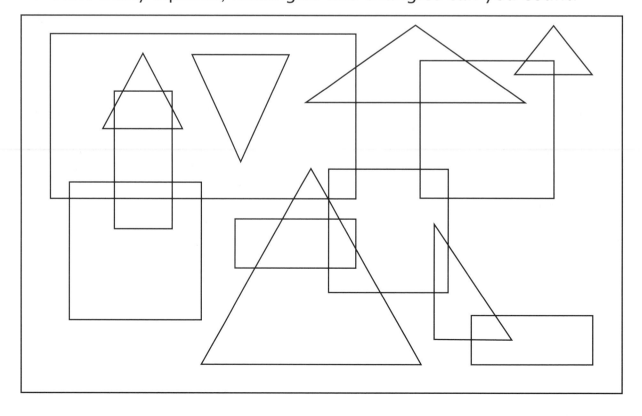

■SCHOLASTIC
www.scholastic.co.uk

Chapter Four

Logic problems

Of the four types of problem included in this book, logic problems are probably the type that young children will find the most challenging to solve. They are required to use quite complex thinking skills to solve logic problems. They have to manipulate each piece of information (or clue) contained in the problem to see what effect they have on each other, and from this, work out the answer. Children who have difficulty solving logic problems may benefit from watching the teacher model how to solve them.

As with any type of problem, children have to be taught to use appropriate strategies to solve logic problems and shown how these methods can be used systematically.

The activities in this chapter teach children the skills and strategies they need to be able to tackle logic problems, including how to:
- read the problem several times before deciding what to do;
- prioritise the information contained in the problem to help them decide on a suitable starting point from which to solve it;
- fix one piece of information and then see what effect this has on the information contained in the rest of the problem;
- select and use practical equipment to help them solve the problem;
- check their answer carefully to make sure that all the criteria of the problem are satisfied.

The activities in this chapter are mainly solved by manipulating practical visual materials, and little recording of data is required.

Wild things

Setting the context
Where the Wild Things Are by Maurice Sendak (Bodley Head) is a story about a little boy called Max who goes on a journey where he encounters different 'wild things'. This problem is introduced using a simple adaptation of this story in which one of the children in the class becomes the central character.

Problem
Daniel (use the name of a child in your class) saw seven wild things in the forest. Some of the wild things had ten horns; the rest of the wild things had five horns. Altogether, Daniel counted 50 horns. How many of each type of wild thing did Daniel see?

Objectives
To solve mathematical problems and puzzles.
To count in tens and fives.
To explain methods and reasoning.

You will need
A copy of photocopiable page 92 for each group; 12 simple 'wild thing' headbands or an enlarged set of the picture cards on photocopiable page 92.

Preparation
Make 12 simple headbands for the children to wear: cut out 18 hand shapes, stick two hands on the front of six of the headbands and one hand on the remaining six headbands.

Solving the problem
● Read the story. Encourage the children to look carefully at the pictures and describe the wild things.
● Choose 12 children to put on the headbands and pretend to be wild things. Tell a simple story about one of the children in the class who goes on a journey and sees lots of different wild things.
● Ask the 'wild things' to stand in a line at the front. Ask the rest of the class to look at them carefully and suggest a way of sorting them into two groups (those with five horns and those with ten horns). Tell the 'wild things' to sit down in their groups.

● Instruct some of the wild things to stand up – for example, three wild things with ten horns and two with five horns. Choose a child to count how many wild things there are standing up, and then sort them into two groups.
● Ask a different child to count how many horns the wild things have altogether. Observe the counting strategy they use. If they count the horns in ones encourage the class to consider if there is a more efficient way of counting the total number. If necessary, demonstrate the strategy (counting in tens and then on in fives (10, 20, 30, 35, 40)).
● Repeat this activity several times. Each time ask a different combination of wild things to stand up. Praise children who use the strategy you have shown them to count the total number of horns.
● Read the problem to the class. Make sure that all the children understand what they have to find out, identifying the given facts that will enable them to solve the problem.
● Show children the wild thing cards and make sure they understand how they can use these cards to help them solve the problem.
● Organise the children into groups of three to work on the problem. Write some similar problems on the board for quick finishers to solve. For example: Daniel saw 11 wild things in the forest. Some had ten horns, the rest had five horns. Altogether, Daniel

counted 80 horns. How many of each type of wild thing did he see?

● Guide children who are having difficulties working out the mathematics, by asking questions such as: *Can you make a group of wild things with a total of 50 horns using only wild things with ten horns? Is this an appropriate solution to the problem?* Agree that it is not a suitable answer because the problem told us that Daniel saw seven wild things and that some of them had five horns and some had ten horns. Show the children how one of the cards can be exchanged for two wild things with five horns. Now ask if this is a reasonable answer to the problem.

Drawing together

● Gather the children together. Choose one group to describe how they solved the first problem. Ask them to illustrate their solution by sticking the appropriate wild thing cards on the board and to explain how they arranged the wild things to make them easier to count.

● Do the rest of the class agree that this is the right answer? How do they know? Does it meet all the requirements of the problem? Demonstrate how the solution could also be recorded as a number sentence $10 + 10 + 10 + 10 + 5 + 5 = 50$

Support

Set a much simpler problem such as: *Daniel saw 25 horns. Using the picture cards show me the wild things Daniel saw.*

Extension

Set more complex problems involving larger numbers or challenge the children to solve problems without using the picture cards.

Further ideas

● Ask the children to make up a wild thing problem of their own for a friend to solve.

● Set a similar problem in a money context, using 5p and 10p coins to make a total.

Moneybags

Setting the context

The giant in 'Jack and the Beanstalk' is very rich – he has gold coins, a hen that lays golden eggs and a magic harp. The giant keeps his gold in three special moneybags. Every evening he tells his wife to bring the moneybags to him. He empties the coins out of the bags and polishes them until they are so shiny that he can see his face in them. Before he puts the coins away again he always counts them very carefully to make sure that none of them have been stolen.

Problem

The giant has ten gold coins. He sorts them into three moneybags:
● The first bag has three fewer gold coins than the second bag.
● The second bag has two more gold coins than the third.
● The second bag has an odd number of coins inside.
How many coins are in each of the moneybags?

Objectives

To solve mathematical problems and puzzles. To understand the vocabulary of comparing and ordering numbers.
To explain methods and reasoning.

You will need

Ten gold coins (these could be chocolate coins or large circles of card covered in gold paper); three moneybags; a copy of photocopiable page 93; ten coins (counters/ £1 coins) for each group.

Preparation

Divide the ten coins between the three moneybags.

Solving the problem

● Show the children the three bags and ask them to guess what might be inside. Tip the gold coins onto the table. Count how many coins there are altogether.
● Ask the children to guess who these 'giant' coins might belong to and then describe how the gold belongs to the giant in

the story, 'Jack and the beanstalk' (see above).

● Now ask one of the children to divide the ten coins between the three moneybags. Count how many coins are in each of the moneybags. Show the children how to record this as a number sentence (such as: $2 + 3 + 5 = 10$).

● Repeat this activity several times. Establish that there are lots of different ways that the giant could split his gold between the three moneybags.

● Read the problem to the children. Tell the children that this type of problem is called a logic problem and that all the information they need to work out the answer is contained in the problem. Ask: *What do you need to find out? What do you know about the number of coins in the first/second/third moneybag? If you add the number of coins in each bag together what will the total be?*

● Demonstrate how you would solve the problem. Work through the problem step by step, explaining everything you do. First, prioritise the information contained in the problem and decide which of the given facts you will use as a starting point. For example, you know that there is an odd number of coins in the second bag, therefore, there must be 1, 3, 5, 7 or 9 coins in this bag. Explore each of these possibilities systematically, starting with the smallest number it could be. Finally, model how you refer back to the original problem to check that your answer meets all the criteria of the problem.

● Tell the children that you would like them to solve a similar problem by themselves. Organise the children to explore the problem in groups of two or three. Give each group a copy of photocopiable page 93, and ten coins/counters to represent the giant's gold.

● Before the children try to solve the problem, remind them to identify the given facts by reading the problem carefully. Observe the different strategies that the children use to solve the problem, such as guessing the answer and then checking it against the problem, or using the equipment to act the problem out. Encourage the children to keep track of what they are doing

and to check their answers by referring back to the original problem.

Drawing together

Gather the class together. Invite several groups to state what they think the answer is, and to explain to the rest of the class the strategies they used to solve the problem. Ask leading questions to help less-able children to structure their explanation: *What was the first thing you did? What did you do next? How did you check that your answer was correct?*

Support
Help the children to understand and then prioritise the given facts and work through the problem practically, using ten coins and three moneybags.

Extension
Challenge more-able children to solve the problem mentally using paper and pencil methods to help them keep track of what they are doing.

Further idea
Set the same problem in other contexts – cakes on plates, flags on sandcastles and spots on ladybirds.

Queen Of Hearts

Setting the context
Say the traditional rhyme:
The Queen of Hearts,
She made some tarts,
All on a summer's day;
The Knave of Hearts,
He stole those tarts,
And took them clean away.

The King of Hearts
Called for the tarts,

And beat the knave full sore;
The Knave of Hearts
Brought back the tarts,
And vowed he'd steal no more.

Problem
The Queen of Hearts made 20 tarts. She arranged the tarts on some plates. Some of the plates had five tarts on and some of the plates had two tarts on. How many plates did she use?

Objectives
To solve mathematical problems and puzzles.
To count in fives and twos.
To explain methods and reasoning.

You will need
Dough; cookie cutters; a large plate; rolling pins (or card, scissors, templates and drawing materials so that the children can make cardboard tarts); paper plates; pencils and paper; two simple cardboard crowns.

Preparation
Make two simple cardboard crowns decorated with heart shapes, and arrange 20 tarts (real/cardboard/dough) on a large plate.

Solving the problem
● Choose three children to act the parts of the Queen of Hearts, King of Hearts and Knave of Hearts. Help them to act out the story while the rest of the class recite the nursery rhyme. Count how many tarts are on the plate altogether.
● Organise the class to work in small groups of three to five. Give each group a rolling pin, some dough, a cutter and a plate. Tell them to make 20 tarts and put them on the plate. (Alternatively, show the children how to draw around a simple template to make the tarts out of card.)
● Ask the children to try and arrange the tarts into piles so that there are the same number of tarts in each pile (such as two piles of ten tarts; four piles of five tarts and so on). Gather the class together and ask the children to demonstrate the solutions they have found using their tarts (10, 5, 4, 2, 1).
● Now read the problem with the children. Ask the children to describe, in their own words, what they think the problem is asking them to do. Help the children to identify the given facts that will help them to solve the problem by asking questions such as: *How many tarts did the Queen of Hearts make? What did she do with the tarts? How many tarts did she put on each plate?*
● Discuss what equipment the children could use to help them work out the answer. Tell the children that you would like them to record the answer. Discuss how they could do this, for example, they could draw tarts

on plates or write a list of numbers.

- Observe the children as they are working. Ensure that the children in each group are working collaboratively. Encourage relevant discussion and dialogue within the groups. Ask the children to check that their answer meets all the criteria of the problem.
- Provide the children with an opportunity to rehearse their explanations and answers before sharing them with the whole class. Ask: *How did you get this answer? How do you know it is right? Is this the only way of solving the problem? How do you know?*
- Pose a second problem for groups that finish quickly: 'The Queen of Hearts made 30 tarts. She arranged the tarts on some plates. Some of the plates had seven tarts on and some of the plates had three tarts on. How many plates did she use?'

Drawing together

- Choose one group to describe, in their own words, what the problem required them to find out, and to demonstrate the strategy they used to solve it. Do all of the groups agree that the Queen of Hearts would have used seven plates?
- Ask different groups to show how they recorded the answer to the problem. Discuss and compare the different methods the children have devised.
- Work through the second problem together. Begin by grouping the biscuits in sevens, note that there are two biscuits left over. Discuss why this arrangement of biscuits is not a suitable solution. Can the children tell you what you must do next?

Support

Ask the children to arrange 18 tarts so that there are the same number on each plate.

Extension

Challenge more-able children to solve the problems mentally using paper and pencil methods to help them keep track of what they are doing.

Further idea

Set the same problem in other contexts, such as spots on dogs, eggs in baskets and so on.

Underwater treasures

Setting the context

Mermaid Megan (or insert the name of a child in your class) is a beautiful mermaid who lives at the bottom of the ocean. Her favourite hobby is collecting shells. She has many precious shells in her collection. She keeps the shells safe in a treasure box. Every day Mermaid Megan takes the shells out of her treasure box to admire their interesting shapes and beautiful patterns. She sorts them into different piles and then puts them carefully back into the box.

Problem

Mermaid Megan has between 16 and 30 shells:
- If she groups the shells in fives there are four left over.
- If she groups the shells in threes there are none left over.
- If she groups the shells in fours there are none left over.

How many shells does the mermaid have altogether?

Objectives

To solve mathematical problems and puzzles. To explain methods and reasoning.

You will need

24 shells placed inside a special box (if you haven't got this many shells, use a mixture of real shells and the shell picture cards on page 94); a copy of photocopiable page 94 for each group; a number line; strips of blue crêpe paper (to represent the sea).

Preparation

Copy, colour, laminate and cut out an enlarged set of shell cards (photocopiable page 94).

Solving the problem

- Ask the children to sit in a circle. Scatter the crêpe paper strips in the circle. Choose one child to be the mermaid. Instruct her to sit in the middle of the circle holding the special box of shells. Tell the class about Mermaid Megan and her collection of beautiful shells. Encourage the mermaid to act out the scene you are describing.
- Pass a few of the shells around the circle so that the children can take a closer look. Ask the children to describe the shells, focusing on the different shapes and patterns. Invite them to suggest different ways that the mermaid might sort her shells such as by size, colour or texture.
- Put the shells back into the special box and read the problem with the children. Tell them that this type of problem is called a logic problem and that all the information they need to be able to work out the answer is contained in the clues. Stress that the clues can be used in any order.
- Ask the children to read the clues again and then say which clue they think it might be sensible to start with and why. Encourage them to see that the first clue will help them to narrow down the number of possible answers. Circle the numbers between 16 and 30 on the number line to show the possible answers.
- Demonstrate how to solve the problem, suggest that you start with the smallest

number of shells that could be in the box. Ask: *What do we need to do to work out if there are 17 shells in the box? What sort of equipment could we use to help us do this?* (Cubes, counters or drawing pictures, for example.)

● Count out 17 shell cards. Ask one of the children to group the shells in fives. Count how many shells are left over. Can the children explain why there cannot be 17 shells in the box? Ask: *Is it necessary to go through the other clues?* Cross the number 17 out on the number line.

● Organise the children to continue working through the problem in pairs. Give each group a copy of photocopiable page 94, scissors, a pencil and paper.

● Observe the children as they work. Do they take a systematic approach to solving the problem (starting with the smallest number)? Do they use a strategy for keeping track of the numbers they have tried?

● Encourage more-able children to see that they can use their knowledge of counting patterns to eliminate some of the numbers. A closer look at the second clue tells us that the answer cannot be a multiple of five. Help the children to notice that there are two multiples of five circled on the number line (20, 25).

Drawing together
● Ask each group to check their answers carefully, to make sure that it satisfies all the criteria of the problem.
● Invite different groups to come to the front and demonstrate the strategy they used to solve the problem. Compare the strategies the children have used. Which method do they think is the best? Why?
● Count the shells in the box to check the answers.

Support
Work alongside the children as they are solving the problem, guiding them towards using appropriate strategies.

Extension
Make the task more challenging by asking the children to solve the problem without using the shell cards.

Further idea
Replicate this problem in a different context, such as a teacher sorting the children in her class into different sized groups, or a child arranging sweets in different piles.

Treasure hunt

Setting the context
Begin the lesson by discussing the children's experience of treasure hunts. Tell the children that you have six pots of treasure and that the treasure in each pot is a different colour. Explain that the children will receive a reward if they can work out, from some simple clues, what colour the treasure in each pot is. The motivation for children to want to solve this problem is the promise of a small reward if they work out the correct answer! The suggested reward here is a sweet. If you would prefer to reward the children with a different treat, the activity can easily be adapted. Instead of putting sweets in the pots, use coloured cubes. Tell the children what reward they will receive if they guess the colour of the cubes in each pot correctly.

Problem
Can you use the clues to work out what colour sweets are in each pot?
- Red is in front of blue.
- Yellow is between blue and green.
- Pink is to the left of orange and red.

Objectives
To solve mathematical problems and puzzles.
To understand positional vocabulary.
To explain methods and reasoning.

You will need
Six small tubs with lids; different coloured sweets such as Smarties (enough for one per child); six Multilink cubes (red, orange, blue,

back

left   right

front

green, yellow, pink); a copy of photocopiable page 95 for each group; scissors; crayons; glue.

Preparation
Sort different coloured sweets into each pot and then arrange the pots on a tray as illustrated in the picture on page 90.

Solving the problem
● Show the children the six pots on the tray. Tell the children that each of the pots has some sweets inside. Explain that there are enough sweets for all of the children to have one – but if they would like a sweet, they must first solve a problem.
● Explain that there are six different colours of sweet. Describe how you have sorted them so that there is a different colour of sweet in each pot. Say that you are going to give the children some clues to help them work out which colour sweets are in which pot and that if they are able to work out the correct answer they may all help you eat the sweets!
● Read the clues with the children. Ask them to describe, in their own words, what you have asked them to find out. Have they seen a problem like this before? Where will they find all the information they need to solve the problem?
● Organise the children to work in groups of two or three. Give each group a copy of photocopiable page 95 and show them how to cut out and colour the six sweet cards to represent the sweets in the pots. Instruct the children, in their groups, to work together to try and solve the problem practically using the cards, the picture of the six pots and the three clues on the photocopiable sheet.
● Talk to the children about what they are doing. Support the children, if necessary, by asking leading questions that help the children to prioritise the given facts, moving them towards using a suitable strategy. For example, encourage them to reason from the information contained in the second clue that the yellow sweets must be in one of the middle two pots. Tell them to put the yellow sweet card on top of

one of these pots on the sheet and then work through the other clues to see what effect this has on the rest of the colours.
● Emphasise to the whole class that because each of the statements in the problem has an effect on each of the other statements, it is important that the children refer back to the original problem to check their answers before they stick the sweets down.

Drawing together
● Gather all the groups together. Choose one group to show their solution and to explain to the rest of the class how they arrived at this answer. Ask: *What was the first thing you did? Why? What did you do next? How did you check your answer?*
● Work with the children to check this group's answer against each of the clues given in the problem. If everyone agrees that this solution meets all the criteria of the problem, then open the pots to confirm whether or not they are correct. If the children do not agree that this solution meets all the requirements of the problem, then ask a different group to show their solution. Continue in this way until an agreed solution is reached.

Support
Work alongside the children as they are solving the problem, modelling appropriate strategies.

Extension
Ask the children to try and write their own logic problem.

Further idea
Replicate the problem in a different context. For example, draw six houses and write a set of clues that the children can use to work out the colour of each front door, or the number of each house.

Wild things

Creative Activities for Maths Problem Solving: Ages 5-7

SCHOLASTIC www.scholastic.co.uk

Moneybags

The giant has ten gold coins.

He sorts them into three moneybags.

The first bag has two fewer gold coins than the third bag.

The second bag has an even number of coins inside.

The third bag has three coins fewer than the second bag.

How many coins are in each of the moneybags?

The giant has 20 gold coins.

He sorts them into three moneybags.

The second bag has fewer than ten coins inside.

The third bag has one more gold coin than the first bag.

The second bag has three more coins than the third bag.

How many coins are in each of the moneybags?

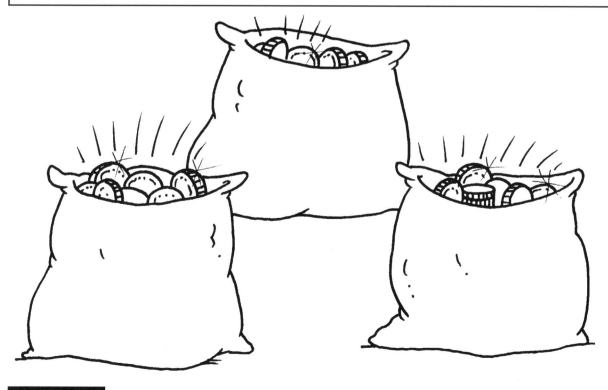

Underwater treasures

The mermaid has between 16 and 30 shells.

If she groups the shells in fives there are four left over.

If she groups the shells in threes there are none left over.

If she groups the shells in fours there are none left over.

How many shells does the mermaid have altogether?

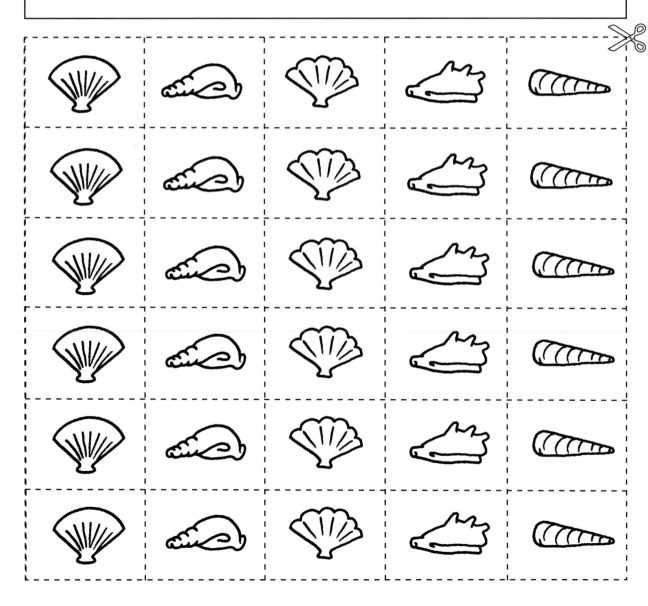

■ SCHOLASTIC
www.scholastic.co.uk

Treasure hunt

- Use the clues to work out what colour sweets are in each pot.

Red is in front of blue.

Yellow is between blue and green.

Pink is to the left of orange and red.

back

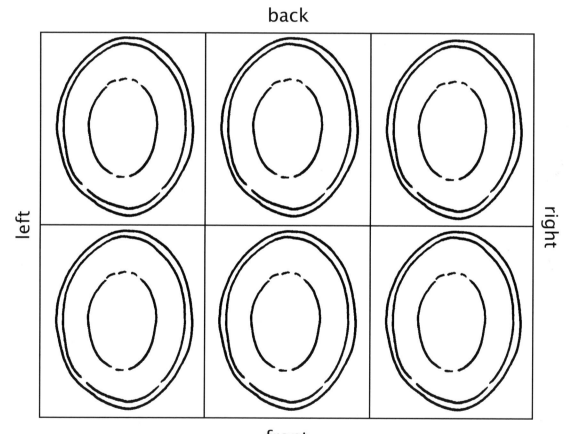

left

right

front

- Cut out and colour the sweet cards. Use them to help you solve the problem.

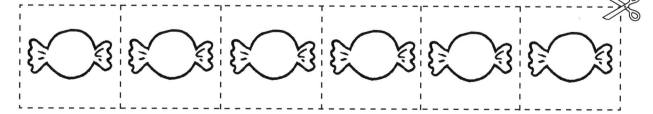

In this series:

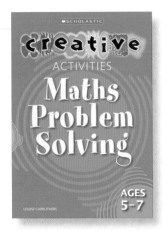

ISBN 0-439-96556-X
ISBN 978-0439-96556-9

ISBN 0-439-96570-5
ISBN 978-0439-96570-5

Also available:

Shortlisted for the
EDUCATIONAL
RESOURCES
AWARDS
2005

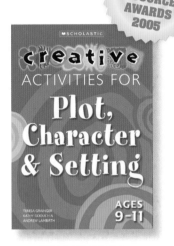

ISBN 0-439-97111-X
ISBN 978-0439-97111-9

ISBN 0-439-97112-8
ISBN 978-0439-97112-6

ISBN 0-439-97113-6
ISBN 978-0439-97113-3

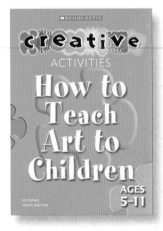

ISBN 0-439-96526-8
ISBN 978-0439-96526-2

ISBN 0-439-96525-X
ISBN 978-0439-96525-5

ISBN 0-439-96524-1
ISBN 978-0439-96524-8

To find out more, call: 0845 603 9091
or visit our website www.scholastic.co.uk